Pelargoniums

Pelargoniums

A GARDENER'S GUIDE TO THE SPECIES AND THEIR CULTIVARS AND HYBRIDS

Diana Miller

B.T. Batsford Ltd, London

First published in Great Britain 1996

© Diana Miller 1996

All rights reserved. No part of this publication
may be reproduced in any form or by any
means, without permission from the publisher

A CIP catalogue record for this book is available
from the British Library
ISBN 0 7134 7283 9

Printed in Hong Kong

For the Publishers
B.T. Batsford Ltd
4 Fitzhardinge Street
London W1H OAH

■ CONTENTS

Foreword by Martyn Rix		6
Acknowledgements		7
List of line drawings		8
List of photographic plates		8
Introduction		10
1	**HISTORY**	15
2	**CLASSIFICATION**	21
3	**INTRODUCTION TO SECTIONS**	30
4	**SYNOPSIS OF SECTIONS**	35
5	**THE *PELARGONIUM* SPECIES**	36
	Campylia (Sweet) de Candolle	36
	Ciconium (Sweet) Harvey	42
	Cortusina (de Candolle) Harvey	55
	Glaucophyllum Harvey	60
	Hoarea (Sweet) de Candolle	64
	Isopetalum (Sweet) de Candolle	70
	Jenkinsonia (Sweet) Harvey	70
	Ligularia Harvey	74
	Myrrhidium de Candolle	87
	Otidia (Sweet) G. Don	90
	Pelargonium de Candolle	95
	Peristera de Candolle	112
	Polyactium de Candolle	118
	Reniformia (Knuth) Dreyer	125
6	**CULTIVATION**	131
7	**PROPAGATION**	137
8	**PESTS, DISEASES AND DISORDERS**	143
	Pests	145
	Diseases	147
	Physiological disorders	148
Glossary		150
Endnotes		156
Select bibliography		158
Species index		160
Plant index		163
General index		172
Sources of information		174

FOREWORD BY MARTYN RIX

Pelargoniums have been popular with gardeners ever since they were first brought to Europe from South Africa during the seventeenth century. Thousands of cultivars have been raised and later discarded in the regal and zonal sections, but the species and some of the early hybrids with scented leaves have retained their popularity for more than three hundred years, kept alive on cottage windowsills, in gardens in mild climates and in the collections of enthusiasts. They are now undergoing a resurgence of interest in their native South Africa.

This book will be used and loved by all growers of pelargoniums. For the first time it brings together into one volume the knowledge of pelargoniums accumulated by botanists and gardeners over the past two hundred years. It incorporates the results of the very latest scientific research published both in South Africa and in Europe, and puts them within the reach of amateurs and professionals.

Diana Miller is especially well-qualified to write this book, being herself a professional botanist and gardener, and for many years a grower, collector and lover of pelargoniums.

Martyn Rix

ACKNOWLEDGEMENTS

Although this book has only one name on its title page, it is the result of the continual accumulation of knowledge by the author as well as that of all the other enthusiasts who have studied and grown the species of this genus for over three hundred years. Without the intrepid collectors of the past, their wealthy patrons and the gardeners who have maintained the resulting plants through the years, we should have no plants to grow, and without the early botanists who meticulously recorded these discoveries, leaving us a legacy of invaluable text, illustrations and herbarium specimens, no records to study.

However, there are many others today whose direct help and advice ensured the completion of this book. The beauty and variety of the species of *Pelargonium* could not be fully appreciated from descriptions alone and my thanks are due to Dick Robinson who has spent many hours photographically recording all these plants. Mrs Hazel Key, the owner of the National Collection of pelargoniums must equally be acknowledged. Her enthusiasm and dedication to the genus has resulted in the accumulation and maintenance of the vast number of species grown in the collection which are available for study, and her encouragement ensured a memorable visit to South Africa to see the plants in their own setting. Alan and Renata Parsley, as well as Fiona Powrie and her colleagues at the National Botanic Garden at Kirstenbosch, enabled us to see a large percentage of the species growing in the wild.

The work of taxonomists around the world, especially those at Stellenbosch University led by Professor van der Walt and Dr Vorster, has resolved many of the taxonomic problems of the genus. Faye Brawner has ensured that I have noted the American point of view and has advised on pests and diseases which are fortunately not known in this country. The assistance of the staff of the Royal Horticultural Society's Lindley Library and of the library and herbarium of the Royal Botanic Gardens, Kew, has enabled me to study older volumes and specimens. Many others from this country, France, USA and South Africa have provided plants or seeds to widen the range of species that I grow.

There are many people to be remembered who have contributed directly or indirectly but who cannot be mentioned individually. Finally I must mention my family who have had to learn to fend for themselves for so many months during the writing of this book.

Illustrations

LINE DRAWINGS

1. The natural distribution of *Pelargonium* in the world
2. The rainfall pattern in southern Africa
3. The natural vegetation of southern Africa
4. A typical fruit of *Pelargonium*. The fruits of related members of the family Geraniaceae are very similar
5. A comparison of a regular flower of *Geranium* with an irregular flower of *Pelargonium*
6. The parts of a typical *Pelargonium*
7. A section through a typical *Pelargonium* flower
8. The geographical areas of southern Africa
9. *P. inquinans* and *P. salmoneum* from Dillenius, *Hortus Elthamensis*
10. *P. zonale* from Commelin, *Icones plantarum horti medici Amstelaedamensis*
11. *P.* 'Graveolens' from L'Héritier, *Geraniologia*
12. Glossary (1)
13. Glossary (2)

LIST OF PHOTOGRAPHIC PLATES

Jacket photographs: front, *P.* 'Schottii'; back, *P. betulinum*

1. *P. ovale*
2. *P. tricolor* near Garcia Pass, South Africa
3. *P. ovale*, *P. tricolor* and *P.* 'Splendide'
4. *P. caylae*
5. *P. frutetorum*
6. *P. inquinans*
7. *P. peltatum* at Worcester Botanical Garden, South Africa
8. *P. quinquelobatum*
9. *P. stenopetalum*
10. Four variations of *P. zonale* grown from seed collected wild in South Africa
11. *P. echinatum* at Worcester Botanical Garden, South Africa
12. *P. xerophyton*
13. *P. grandiflorum* from Andrews, *The botanist's repository*
14. *P. laevigatum* near Oudtshoorn, South Africa
15. *P. lanceolatum*
16. *P. tabulare* at Kirstenbosch Botanical Gardens, South Africa
17. *P. longifolium* at Worcester Botanical Garden, South Africa
18. *P. pinnatum*
19. *P. rapaceum*
20. *P. cotyledonis*
21. *P. endlicherianum*
22. *P. tetragonum* near Anysberg, South Africa
23. *P. quercetorum*
24. *P. alpinum*
25. *P. divisifolium*
26. *P. rodneyanum*
27. *P. spinosum* at Worcester Botanical Garden, South Africa
28. *P. stipulaceum* at Worcester Botanical Garden, South Africa
29. *P. trifidum*
30. *P. longicaule* var. *longicaule*
31. *P. suburbanum* subsp. *suburbanum*
32. *P. dasyphyllum*
33. *P. betulinum*, Cape of Good Hope, South Africa
34. *P. citronellum*
35. *cordifolium*
36. *P. cucullatum* subsp. *cucullatum* near Rooeils, South Africa
37. *P. cucullatum* subsp. *tabulare* at Miller's Point, South Africa
38. 'Graveolens'; *P. radens* and *P. graveolens*

39 *P. hermanniifolium* from Villersdorp, South Africa
40 *P. quercifolium* from plant collected in South Africa and grown from seed from South Africa; *P. panduriforme P. pseudoglutinosum* and *P.* 'Royal Oak'
41 *P. sublignosum*
42 *P. australe*
43 *P. iocastum*
44 *P. bowkeri*
45 *P. caffrum*
46 *P. gibbosum*
47 *P. luridum* from East Transvaal, South Africa
48 *P. multiradiatum*
49 *P. schizopetalum*
50 *P. exstipulatum*
51 *P. fragans* and *P. odoratissimum*
52 *P. ionidiflorum*
53 *P. reniforme*
54 *P. sidoides*

Line drawings: figures 1, 2, 3, 4, 12 and 13 are by Duncan Lowe; 5, 6, 7 and 8 by Valerie Price; 9, 10 and 11 are by courtesy of the Lindley Library of the Royal Horticultural Society.

All photographic plates from the Harry Smith photographic collection except 23 by A.D. Schilling, 47 by Susyn Andrews, and 2, 7, 11, 14, 16, 17, 22, 27, 28, 31, 33, 36, 37, 39 and 43 by the author. Plate 13 by courtesy of Hazel Key.

■ INTRODUCTION

There are few gardeners who have not at some time grown a zonal, a regal or an ivy-leaved pelargonium, all of which were as popular with the Victorians as they are today. These plants comprise one of the economically important groups in the world of commercial horticulture. They are the result of nearly two hundred years of breeding, initially in Europe and not long after in America, Australia and elsewhere. However, all these thousands of cultivars were derived originally from no more than a handful of species which may be found growing wild in South Africa. Some enthusiasts grow a selection of scented-leaved cultivars or perhaps of some other selected groups within the genus such as the Uniques, but even the inclusion of these only begins to cover a small proportion of species that are waiting to be seen.

The genus *Pelargonium* includes well over two hundred distinct species. It must be almost unequalled in the plant kingdom for the enormous diversity of form exhibited by its members. The choice is vast including shrubs, annuals, succulents and tuberous plants; plants of only a few centimetres high and others reaching several metres; plants with fragrant flowers or with scented foliage, some of which are used in the commercial production of aromatic oils; plants which have spines and others which have been valued for their medicinal properties. To restrict oneself to the well-known zonals or regals is to lose the pleasure of experiencing a whole world of exciting and

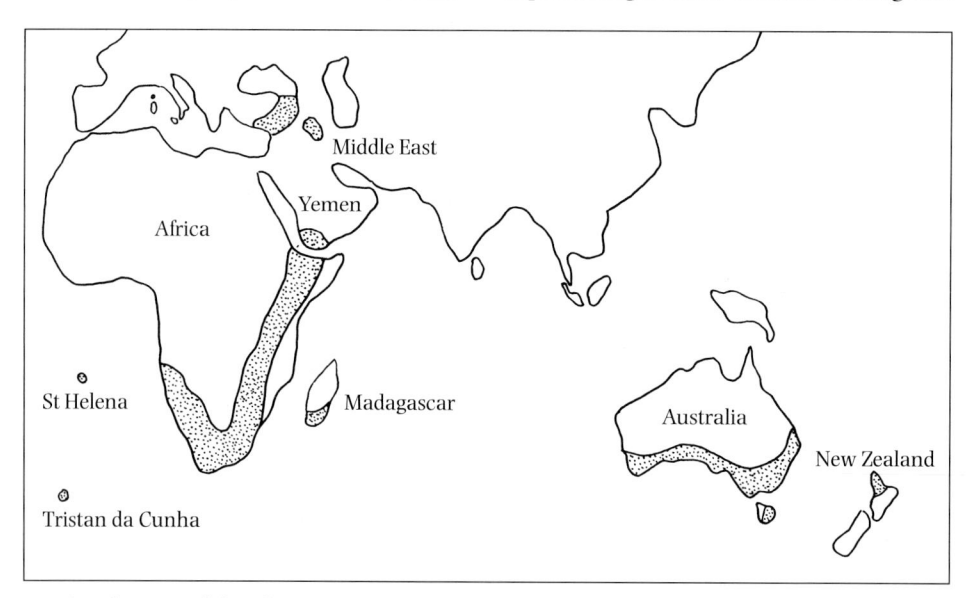

Fig. 1. The natural distribution of *Pelargonium* in the world

fascinating plants which are not difficult to grow and far less prone to pests and diseases than their pampered, highly-bred relatives. Indeed perhaps the reintroduction of a species into the breeding line of some of these highly developed cultivars may reintroduce a degree of natural resistance to certain diseases which can cause problems to growers.

The majority of the species of *Pelargonium* are to be found in southern Africa and of these, over 80% are concentrated in the south-western corner in the area of winter rainfall. About twenty grow in East Africa and about a dozen species are to be found wild outside Africa, some in Australia, one or two in New Zealand, two in the Middle East and one or two on the islands of Madagascar in the Indian Ocean and St Helena and Tristan da Cunha in the southern Atlantic (fig. 1). All these localities are characterized by relatively low rainfall which in many cases falls mainly in one season of the year (fig. 2). No species grow in areas of high humidity and those that are found nearer the equator live at higher altitudes where the temperatures are not excessive. Most species therefore do not have any special cultural requirements as long as the plants are grown in frost-free conditions, in a well-drained soil and without excess humidity. In their native habitat, some experience short periods of frost or snow but with low humidity and very well-drained soils.

This large genus is sub-divided into sections. The divisions of the species into sections on morphological and other features, closely relates, in most cases, to areas of similar climatic conditions with a characteristic type of vegetation, often within one relatively small geographical

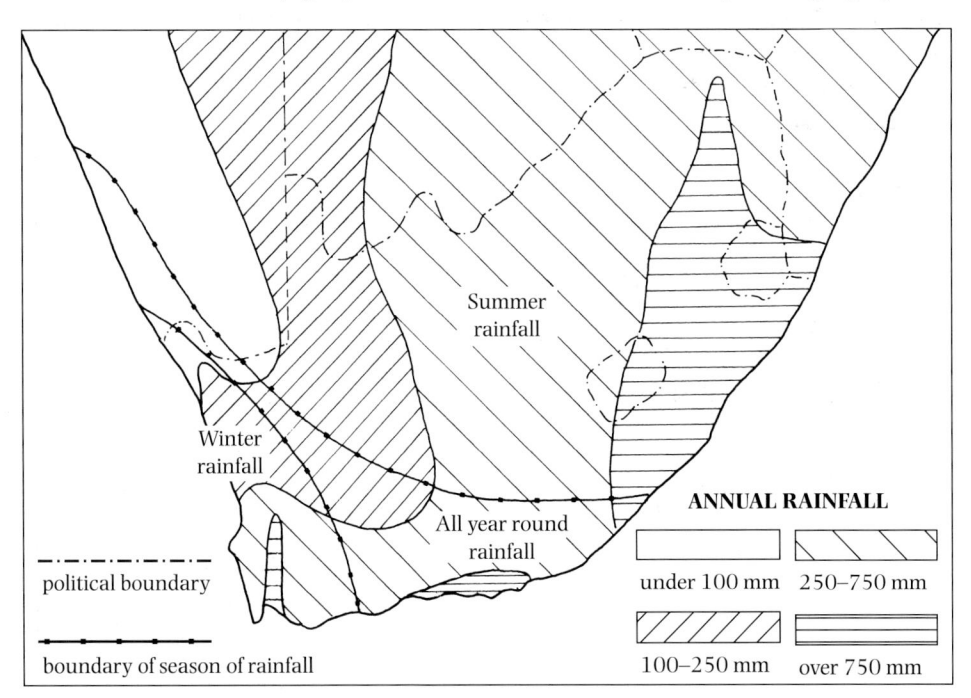

Fig. 2. The rainfall pattern in southern Africa

area (fig. 3). For example all members of the section *Cortusina* are to be found in southern Namibia and the north-western part of Cape Province in desert or semi-desert regions of very low annual winter rainfall. Here the rainfall may be as little as two or three centimetres a year but plants may be found sheltering between rocks or in loose sand. On the other hand, the members of the section *Reniformia* are to be found towards the southern and eastern part of South Africa in areas of summer rainfall. Some species are found in long grass, others at the edges of forest in regions with ten times as much rain.

The vast number of different species found in the south-western part of Cape Province are associated with a Mediterranean type of climate and vegetation which explains why the genus as a whole thrives in southern Europe and California, both having similar environmental conditions. The adaptation of each species to the microclimate in which it is found, explains the enormous variation in habit. Even within a single species in the wild, the variation in form of the species collected from different localities can be quite dramatic. One of the succulent species growing in sand in full sun might be stunted with a very swollen stem but if growing with some protection under a shrub, might become quite tall but elongated. In the more regulated conditions of cultivation these variations are less marked. In some cases the differences between cultivated and wild plants may be sufficient to cause complications in the identification of the species.

There are remarkably few hybrids to be found in the wild. This is perhaps because each species has its own ecological niche

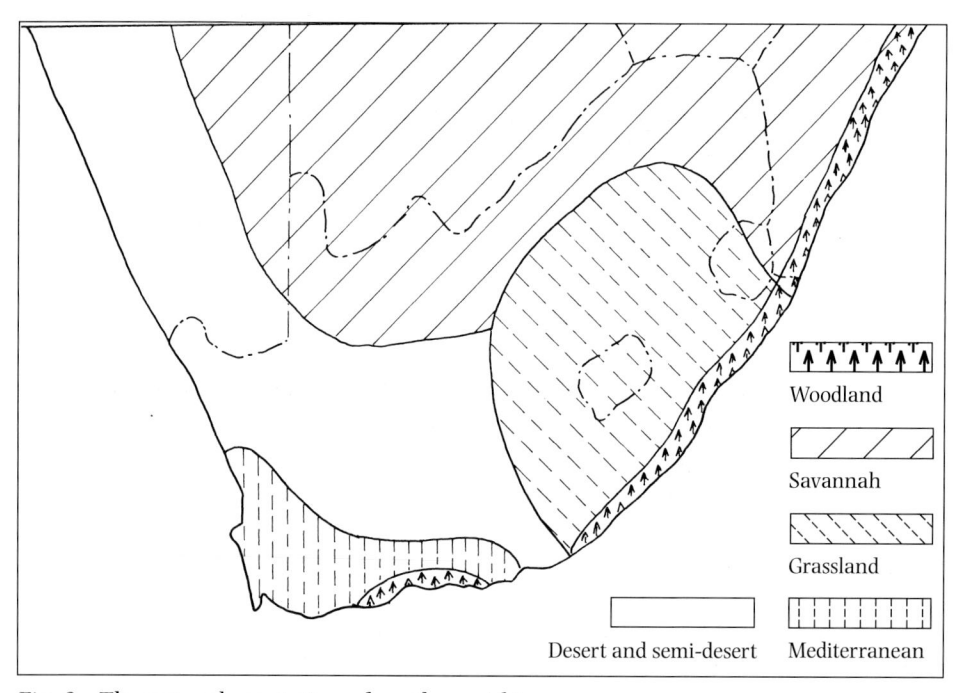

Fig. 3. The natural vegetation of southern Africa

and different species within a section are not often found in close proximity. Sometimes natural hybrids do occur, such as that between *P. cucullatum* and *P. betulinum*, both in the section *Pelargonium*, where two species grow near to one another. However, even when growing close together, hybrids between species of different sections are even more uncommon because there is little compatibility between them and their differing chromosome numbers, although crosses between species of closely related sections do occasionally occur. There are endless possibilities for the gardener to create new and exciting hybrids between species within a section, or to experiment with species of different sections. If there is no other reason, perhaps this is an excuse for more species to be grown so that new and unique plants may be created for the gardeners of the future.

This book includes some of the cultivars derived from the species but does not attempt to describe the highly-bred modern cultivars nor to include specific details of growing or exhibiting the zonal or regal cultivars. There are many other publications available, written by those who are the real experts in the production of plants for the show bench. It does, however, aim to introduce the species to all those who are interested in growing less usual plants and to all enthusiasts of the genus. There are a few species that have not been included, one or two that are very small and weedy, but the rest mainly because they are not in cultivation or are so rare as to be almost impossible to obtain. In these cases any descriptions could not have been prepared from live material and would have been copies of other work.

The bulk of the book consists of descriptions of the species arranged alphabetically with related plants within each botanical section. A table is provided to help identify the section required. A key would be ideal but unfortunately this is almost impossible using the morphological characteristics alone as there are so many exceptions within the species. The limits of the sections have been confirmed by modern botanical techniques but these are mostly unavailable to the average enthusiast. The taxonomic and historical details of each section, its geographical centre and botanical characteristics is followed by a key to identify an unknown species. The scientific names are used throughout, not only to be botanically correct and reduce confusion but also because very few have a widely used vernacular name. I have, however, included the meaning of the botanical names. This may make a name easier to remember for those unused to using latin, because once translated and with a reason, the name often makes more sense. Those unfamiliar with botanical convention, may be confused by the name following the latin name. This is the authority for the name, the person responsible for the first publication of that name, which is explained in more detail in the chapter on classification.

I have also added abbreviated references to illustrations as, although it would be have been ideal, it was not possible to include nearly two hundred colour photographs in this book. There are other books, both old and new, which also have pictures of the species, but just a few standard references have been chosen here for simplicity. The bibliography will give full details of these and many others. Each description has been standardized as far as is practical for a direct comparison, using the features needed to identify the plant. It does not include detailed characteristics of features such as the types of hairs or glands which are of greater interest to the taxonomist, nor does it include details of features which are not markedly different from one species to another.

The botanical description is followed by brief notes on any characters which might give a quick guide and distinguish the plant from a closely-related species. Information about the discovery and history of each species follows as well as any specific cultivation notes and an attempt is made to resolve some of the confusion in the nomenclature.

I hope that this book will introduce others to the fascinating world of the species of genus *Pelargonium* and that they too may enjoy growing, and naming, these endlessly variable plants.

1

History

The discovery and introduction of the first *Pelargonium* species into the great gardens of Europe is closely connected with the history of the opening up of trade routes to the East, and later with the explorations into the interior of southern Africa.

After the famous explorers had discovered the route around the southern tip of Africa and the spice trade was established, the British East India Company was chartered in 1600 to create permanent trading posts. Two years later a similar company was set up by the Dutch. The ships passing around the Cape of Good Hope had regular stopping places for water and fresh food near the Cape and further east at Mossel Bay and Algoa Bay. Soon permanent settlements were created by the sheltered harbour near present day Cape Town. A garden was created to grow fresh fruit and vegetables to replenish passing ships, so the interest in plants began. Ships carried ship's surgeons many of whom would also have been naturalists, concerned with all aspects of wild-life especially new and exciting plants either to take home or for use as food for the sailors. At Leiden in Holland, one of the earliest botanic gardens was established in 1577 just as trade was beginning to flourish, and all ships which left Dutch ports had orders to bring back seeds from all over the known world. As a result, about six thousand different plants were grown in the botanic garden in the early part of the eighteenth century.

The earliest plants from South Africa were bulbs and the first recorded *Pelargonium* was the tuberous-rooted, *P. triste*. It is not known whether or not this was brought first to Leiden but the possibility exists that it was introduced there before 1600. It certainly grows wild in the regions near the early settlements on the coast near present day Cape Town, where ships would have anchored. Other pelargoniums in similar areas had much more showy flowers but *P. triste* has a tuber which would have been more able to survive the long journey back, and also it may have been collected as a possible food source or simply for the night-scented flowers. Whatever the reasons, it somehow arrived in France. We know this because the British gardener John Tradescant meticulously recorded all his transactions, and in one account he notes that he obtained this species with a hamper of seeds from his friend René Morin in Paris in 1631. Even then, its true origin was unknown as it was for many years considered to be a native of India, presumably because it was brought back on a ship returning from the East.

Paul Hermann was perhaps the first serious botanist to visit the region and he was able to collect and press examples of the South African flora while on his way to Ceylon (now Sri Lanka) in 1672. Many of his herbarium specimens were sent back from the Cape, some of which still survive today together with several useful publications. Seeds were sent to Jacob Breyne who illustrated one or two *Pelargonium* species in 1678. On his return to Holland, Hermann was appointed Professor of Botany and dedicated his time to the gardens in Leiden. His first catalogue lists nine pelargoniums, all of which are known today. Jan Commelin, Director of the newer botanic garden in Amsterdam, also received plant material and two years later was listing one or two new species. After he died in 1698, his work was completed by his nephew, Caspar Commelin. By 1710, about ten more new species had been added to this collection.

Though there were not many botanists at this time there was a tremendous exchange of knowledge, especially between the countries of northern Europe, so that any new plants were quickly passed between the botanic gardens such as Leiden, Amsterdam, Oxford and the physic garden of Chelsea, as well as between the amateur botanists and the nobility. There were also no major wars or political disputes between these countries and their scientists appear to have visted one another quite freely. They also seem to have been able to work in countries other than their native land with no difficulty and hence even greater exchange of information and experience was gained. There was a close relationship between Holland and England, and William of Orange appears to have been very interested in new plants. When, in 1689, he acceded to the throne of England, he brought many of his Dutch compatriots with him. One of these was William Bentinck, the head steward, later to become the first Earl of Portland. He was also superintendant of the royal gardens and was known in the British pelargonium world as the introducer of *P. cucullatum* in 1690.

The interest in plants from far away places was intense and pelargoniums came into gardens with many of the other exotic species from the Cape. As well as botanic and royal gardens, there were several interested and wealthy amateurs who became patrons of plant collectors and developed gardens of their own. One of these was James Sherard, an apothecary with a large estate at Eltham in Kent, at which he created a garden noted for its rare plants, employing the German born botanist Johann Dillenius. Dillenius was to record several important pelargoniums in his *Hortus Elthamensis* in 1732 as well as noting the different morphology of the flowers of the African species which were at the time classified with geraniums. At about the same time the Bishop of London, Henry Compton, kept one of the famous contemporary gardens at Fulham which included some pelargoniums.

Back in South Africa, explorations inland to the north and east resulted in the discovery of more new species while the garden of the Dutch East India Company was expanding to grow new medicinal herbs and plants. One of the master gardeners was Heinrich Oldenland, a pupil of Hermann, who could have been responsible for sending *Pelargonium* seeds to Leiden. Simon van der Stel and later his son, were governors of the province during this era of expanding interest in exploration and plant collecting at the end of the seventeenth century. They are known to be responsible for the dispatch of seed of *Pelargonium* species to Holland such as *P. zonale*. There are also records of consignments of chests of plants for both William of Orange and the medicinal gardens of Amsterdam.

Another important name in the history of the *Pelargonium*, is Johannes Burman, successor to Caspar Commelin as professor of Botany in Amsterdam. It was Burman who first used the name *Pelargonium* in 1738, for some of the African species of *Geranium*. Not long after this, Linnaeus published his *Species Plantarum* which although it established the binomial system of naming, did not recognize *Pelargonium*. Such was the stature of Linnaeus at the time that it was to be another forty years before the name was finally approved and even Burman's son Nicolaas did not have the courage to follow his father's opinion.

During the second half of the eighteenth century, there were several botanists who published remarkable works which included pelargoniums. In 1787, Antonio Cavanilles from Madrid, illustrated about seventy in black and white. Meanwhile in Austria, Nikolaus Jacquin, a botanist and artist, and later a professor at Vienna, was responsible for a number of beautifully coloured volumes illustrating the plants grown at the time, especially in Schönbrunn palace. This was the summer residence near Vienna of the Emperor of Austria whose parterres were designed and laid out by gardeners from Leiden. The garden contained one of the most important plant collections of the era, acknowledged as a rival to Kew. In 1780, during a cold winter, the greenhouse stoves were not lit with the loss of most of the tender plants.

As a result, new collecting expeditions were sent by Emperor Joseph II to replenish the gardens. Two collectors, Franz Boos and Georg Scholl travelled to the Cape and many of Jacquin's illustrations were a result of their collections. Scholl was to remain there for ten years.

Arguably, the most important taxonomic works were the illustrations and unpublished descriptions of Charles-Louis L'Héritier, a French magistrate and botanist. Unfortunately, he was assassinated in Paris before some of his work was published, but in it he had described nearly ninety species, most of which were used in Aiton's *Hortus Kewensis* in 1789. He is known to have corresponded with Cavanilles and other contempory botanists and worked in England where he had a serious misunderstanding with the eminent Sir Joseph Banks.

By this time in South Africa, several well-trodden, though hazardous, tracks were in use towards Mossel Bay and George on the southern coast, Oudtshoorn and Willowmore inland to the east as well as routes to the north-west. These were to be gradually explored and extended over the coming years. In 1772 a Scottish gardener, Francis Masson arrived at the Cape of Good Hope from Kew Gardens with orders to collect seeds and plants. Over the next twenty years, he made a number of journeys to the north and east of the Cape collecting hundreds of plants, many of which were the new species of *Pelargonium* listed in Aiton's *Hortus Kewensis* and illustrated by L'Héritier. During his various expeditions, he met and travelled with several other collectors including Scholl, Oldenburg and Carl Thunberg, a pupil of Linnaeus from Sweden. With so many different nationalities it is interesting to speculate on the language in which they conversed. Thunberg however was on his way to collect plants in Japan but at this time, Japan was only accessible to the Dutch. He therefore was sent to the Cape of Good Hope to learn the language and to work with the Dutch East India Company before travelling further East. During this time he was able to collect many plants, including pelargoniums.

Another Kew gardener was Anton Pantaleon Hoveau, from Warsaw, whose name was later anglicized to Anthony Hove. On his way to India in 1795, he collected seventeen pelargoniums from near the Bay of Angra, on the south-western coast of Africa, about five hundred miles north of the Cape. Unfortunately only three survived, *P. crassicaule*, *P. cortusifolium* and *P. ceratophyllum*, because apparently the rest 'did not like the accommodation on the Man of War'. It is a pity that we shall never know what else was collected in the extreme conditions he described where the heat blistered the feet and the plants grew in chasms of white marble rock.

As well as botanists and gardeners, plants were collected by explorers such as William Paterson who was sent out by the Countess of Strathmore in 1777. He joined forces with the Dutchman, Captain Robert Gordon to travel northward along the coast where they discovered *P. sibthorpifolium* and *P. klinghardtense* as well as the Orange River.

At about this time, in the 1780s, there were at least three major nurseries in London selling pelargoniums from the Cape. James Colvill founded a nursery in the Kings Road, Chelsea, specializing in Cape bulbs which was to become a famous institution and eventually housed a very large collection of pelargoniums. It was from working as a foreman for this nursery from 1819 that Robert Sweet was later able to produce his five volumes of Geraniaceae. A rival of Colvill was Andrew Henderson at Pine Apple Place in the Edgeware Road and another, the Vineyard Nursery of Lee and Kennedy based at Hammersmith. In the 1780s, the Vineyard was known for its new Australian plants which were just being discovered following the expeditions of Captain Cook and Joseph Banks. Plants such as *P. australe* found in 1792 were propagated by organizations such as this who sent plant collectors to anywhere thought to be profitable enough to satisfy the demands of the wealthy owners of large estates for new plants to fill their great greenhouses. The Vineyard was even known to supply plants for the Empress Josephine at Malmaison in France, despite the political unrest. The artist Henry Andrews was later to marry the daughter of John Kennedy and so had access to a wealth of material for his paintings of pelargoniums and other plants.

The era of relative calm on the political scene was now changing as the turmoil caused by French revolution affected Holland and neighbouring countries. In 1793, Holland was overrun by the French and the British were authorized to take control in South Africa to protect the area from the French. Ten years later war resulted between the British and Dutch for control, but the Dutch influence had waned by this time and the British thereby became more influential in the pelargonium world. However a large proportion of the species known today had already been collected and named by this time and there was still a tremendous exchange of plants between European countries.

Plant collecting in the region had became more risky but in England, there was great interest in the new hybrids which were appearing. Initially, the process was not understood and hybrids occurred as the result of accidental crosses. Later the only certain parent was the female from which the seed was collected whereas the male was surmised by guesswork. However by the early part of the nineteenth century, gardeners were obviously aware of the process and were beginning to make deliberate crosses. Andrews in his *Geraniums* of 1805–1806 as well as in *The botanist's repository*, begins to suggest the idea of the difference between the types which have arisen in the garden, and the species from Africa. Sweet a few years later was already suggesting parentages, sometimes with great authority as if he was aware of which were the result of deliberate pollination and which were accidental. He also bemoans the fact that the old species were becoming rare and no longer widely cultivated because of the greater preoccupation with the new hybrids which were appearing. He adds that another reason was the pursuit of the new Cape Heaths but that he was hoping to reverse the fashion by the publication of his volumes of *Geraniaceae*. There were still the amateur enthusiasts for the genus, such as Robert Jenkinson, treasurer to the Horticultural Society, and cultivator of 'geraniums', and Sir Richard Colt Hoare of the famous Stourhead gardens, after whom Sweet named two sections, but even these were growing and raising hybrids as well as species.

The very early precursors of the zonal and regal cultivars began during this period and by the middle of the century there was a tremendous interest in both types. Plants with scented foliage, the uniques and the ivy-leaved cultivars as well as zonals with coloured leaves became equally sought after. Fashions went through cycles but in general the aim was to develop perfectly round, brightly-coloured, large and showy flowers. The majority of the species were still retained as curiosities.

Pelargoniums were introduced into America early in their history and spread across the continent as it was colonized, eventually reaching California where the climate was ideal for their growth. The popularity of the plants in Europe, their ease of cultivation and their bright colours must have been reminders of an earlier life, and perhaps one of the reasons why they were taken across the Atlantic and later to Australia.

During this century there has been an explosion in production of new cultivars in all the main groups of *Pelargonium* in many parts of the world. As a result of the two world wars, hundreds of the older cultivars were lost as the big gardens of Europe became derelict when their gardeners left for the battle lines. Even if the plants survived, their names were mislaid or muddled. After the second world war, the interest gradually revived in both species and cultivars in many parts of the world. More recently, seed-raised cultivars for bedding have been developed. Also for this market, the use of meristem culture for the propagation of enormous quantities of identical plants has become a very profitable industry. Unfortunately, instead of thousands of different cultivars, the trend is for the mass production of a few selected, highly-promoted cultivars so that the selection is limited by commercial gain. However, there are always the traditional nurseries and the dedicated amateurs who continue to conserve the enormous number of old and new cultivars, as well as the hybridizers who persevere in their search for new and improved hybrids. This is why the species are so important as the input of new characteristics may produce exciting and unexpected variations.

2

Classification

The genus *Pelargonium* is classified in the family *Geraniaceae*. Several genera of very limited horticultural value were once included in this family but have now been reclassified by most taxonomists into other families. Five remaining genera, closely related to *Pelargonium*, are of great interest to the gardener. They may be distinguished by the diagram below.

All of these five genera have a similar elongated fruit with five individual sections, known as mericarps, each containing one seed (fig. 4). When these are mature, they split apart and the plumed seed is released. *Monsonia* is found mainly in Africa with a few species in Asia. It was named after Lady Anne Monson, a descendant of Charles II.

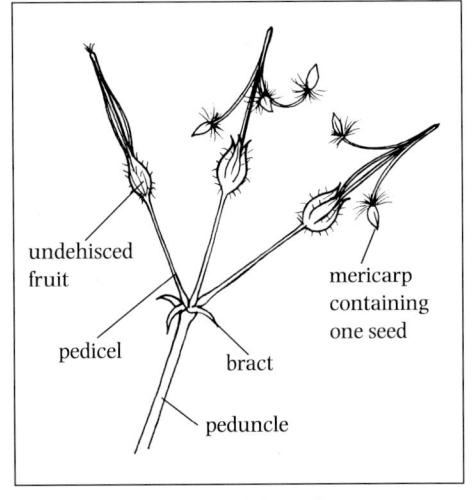

Fig. 4. A typical fruit of the *Pelargonium*. The fruit of related members of the family Geraniaceae are very similar

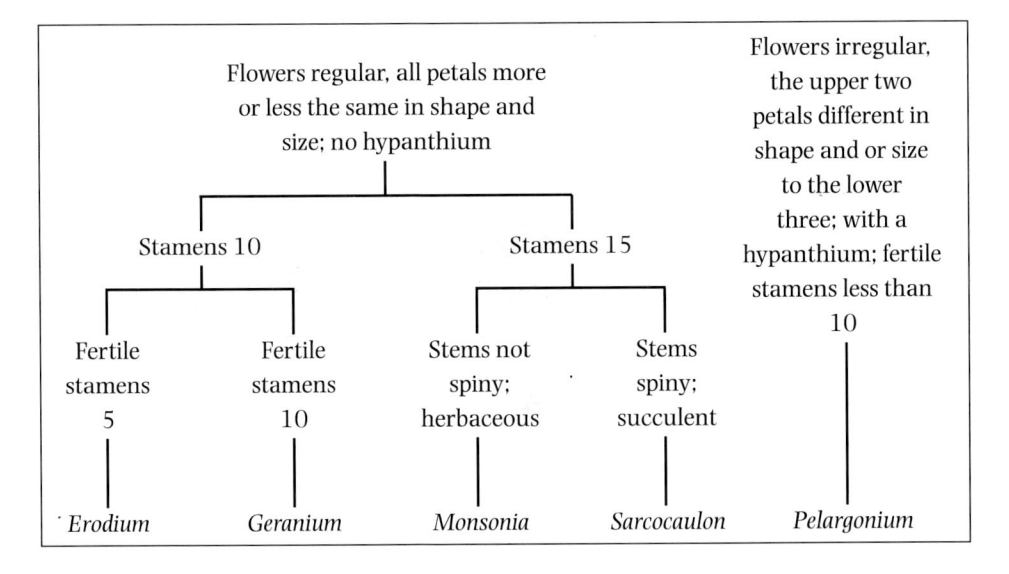

```
                    Flowers regular, all petals more                Flowers irregular,
                    or less the same in shape and                   the upper two
                          size; no hypanthium                       petals different in
                                   |                                shape and or size
                 ┌─────────────────┴─────────────────┐              to the lower
           Stamens 10                          Stamens 15           three; with a
                 |                                   |              hypanthium; fertile
           ┌─────┴─────┐                   ┌─────────┴─────────┐    stamens less than
        Fertile     Fertile             Stems not            Stems  10
        stamens     stamens             spiny;               spiny;  |
           5           10               herbaceous           succulent
           |           |                    |                    |     |
        Erodium     Geranium             Monsonia           Sarcocaulon  Pelargonium
```

Sarcocaulon is limited to southern Africa where it is called Bushman's candle. The name is aptly taken from the Greek meaning fleshy stem. The remaining three genera were included by Linnaeus in the one genus *Geranium*, named from the resemblance of the fruit to a crane's bill, and crane's bill is still used as a vernacular name for some species today. The true*Geranium* has a wide distribution through-out the temperate regions of the world. *Erodium* species were separated and given the name after the Greek for heron, again for the fruit-shape, and these species occur mainly around the Mediterranean regions. Yet another name had to be found for the third genus and the greek for stork, 'pelargos' was chosen, hence *Pelargonium*. The differences between the regular flower of a *Geranium* with five equal petals and ten fertile stamens and the irregular flower of a *Pelargonium* with the upper two and lower three petals distinct are compared in fig. 5.

The great majority of *Pelargonium* species are found in South Africa with a few in Australia, eastern Africa, the Middle East and some islands including St. Helena and Madagascar. There should be no cause for misidentification between the genera as each are easily distinguished from one another, but the confusion that has arisen over the nomenclature of *Pelargonium* and *Geranium* should be mentioned. As explained previously, all pelargoniums were initially included in the genus *Geranium*, but although it is two hundred years since they were separated, the name Geranium has become re-established as a vernacular name for pelargoniums and especially for the zonal cultivars so widely used in bedding. There was perhaps some excuse for the complaints by Andrews in the introduction to his first volume of *Geraniums* in 1805 when he refused to adopt the new generic name:

> This genus has already sustained two innovations; which if we were to adopt, we should have six more to make in addition; being already in possession of that number of an equally distinct character with the pelargonium and Erodium (the innovating genera of M. L'Héritier, a late French botanist), the former having seven fertile tips; the latter five

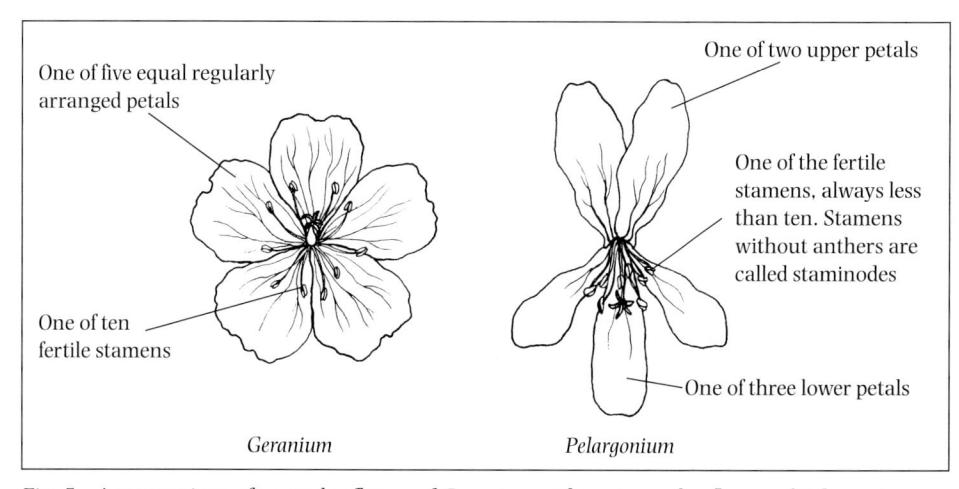

Fig. 5. A comparison of a regular flower of *Geranium* with an irregular flower of *Pelargonium*

only fertile, and these not always permanent: so that if such generic divisions were generally adopted, the approach to botanic science would be so choked up with ill-shaped, useless lumber, that, like a castle in a fairy-tale, guarded by hideous dwarfs, none but a Botanic Quixotte would attempt investigation.

It was after all, only a few years after L'Héritier's publication. However, although Andrews admits that he was in the minority, he really was adamant that the classification of Linnaeus should not be altered and makes scornful comments to this effect throughout his texts:

> To dissent from most of the late publications on the science of botany, may seem presuming; yet when it shall be considered, that we take Linnaeus for our sole guide.... We must therefore after him, think no generic division necessary.... M. L'Héritier and his followers must, therefore, forgive us for ranking one of his pelargoniums under the old-fashioned standard.

There was obviously rivalry between botanists then as there is today in controversial taxonomic issues, and sometimes a decision can only be a personal opinion. There is, however, no excuse today for the incorrect use of the name 'Geranium'. No one still calls an *Erodium*, a geranium. However, even in the *Gardener's Chronicle* of 2 March 1901, there is a note indicating irritation with the inaccurate use of the names: 'In spite of repeated correction, and even in spite of a Pelargonium Society, now defunct, the average gardener and naturally the average layman, speaks of Geraniums when he means Pelargoniums.'

The characteristics of a typical *Pelargonium* are illustrated in fig. 6 and fig. 7.

Pelargoniums may be woody, succulent or herbaceous, evergreen or deciduous, and some have tubers. All have alternate petiolate leaves with stipules. The leaf venation may be palmate or pinnate, simple or compound and of almost any shape. In some sections, the petioles or the stipules or both are persistent and remain on the stems for several seasons. These may harden to form spines which presumably have a protective function for the plant. This loss of the lamina while the petiole is retained is a somewhat unusual habit in the plant kingdom. Another feature which also must have some protective value for the plant is the presence of aromatic oils in the leaves of many species; these are also a useful tool for identification.

The flowers are arranged in umbels with all the flower stalks emerging from one point at the end of the peduncle. In a true umbel, the youngest flowers are at the centre of the flower head but in pelargoniums, the oldest flowers are at the centre, hence the inflorescence is sometimes termed a pseudoumbel. Each flower has five sepals with the posterior one, that is the uppermost, modified so that a nectary is formed at the base. The length of this nectary is characteristic for each species and in some is a mere indentation but in others may be several centimetres long. It is fused to the pedicel and known as the hypanthium. At the end of the hypanthium there is a hump, sometimes quite pronounced, and the relative lengths of the hypanthium and pedicel are often important in the classification of species. The five free petals are often clawed and arranged so that the upper two are separated from the lower three, and there is usually a distinct difference in size, shape and markings. This distinction is less obvious in the modern zonal and regal cultivars which were deliberately bred in Victorian times to bear regular flowers with rounded petals.

The arrangement and positioning of the petals, together with the differing lengths of pedicel and hypanthium, create the wide range of flower formation to be found within the genus as each has evolved alongside the appropriate pollinating insect found in its native habitat. Of the ten stamens, seven or fewer bear fertile pollen and this number is often another significant feature in the recognition of a section. The remaining infertile stamens are known as staminodes which are sometimes curved or developed in some other characteristic manner. The filaments are joined, occasionally for most of their length. The style divides into five stigmas which open and become receptive to pollen after the anthers of that particular flower have dehisced, thus avoiding self-pollination. Once the flower is fertilized, the mericarps swell and the fruit elongates to form the typical shape from which the generic name was chosen.

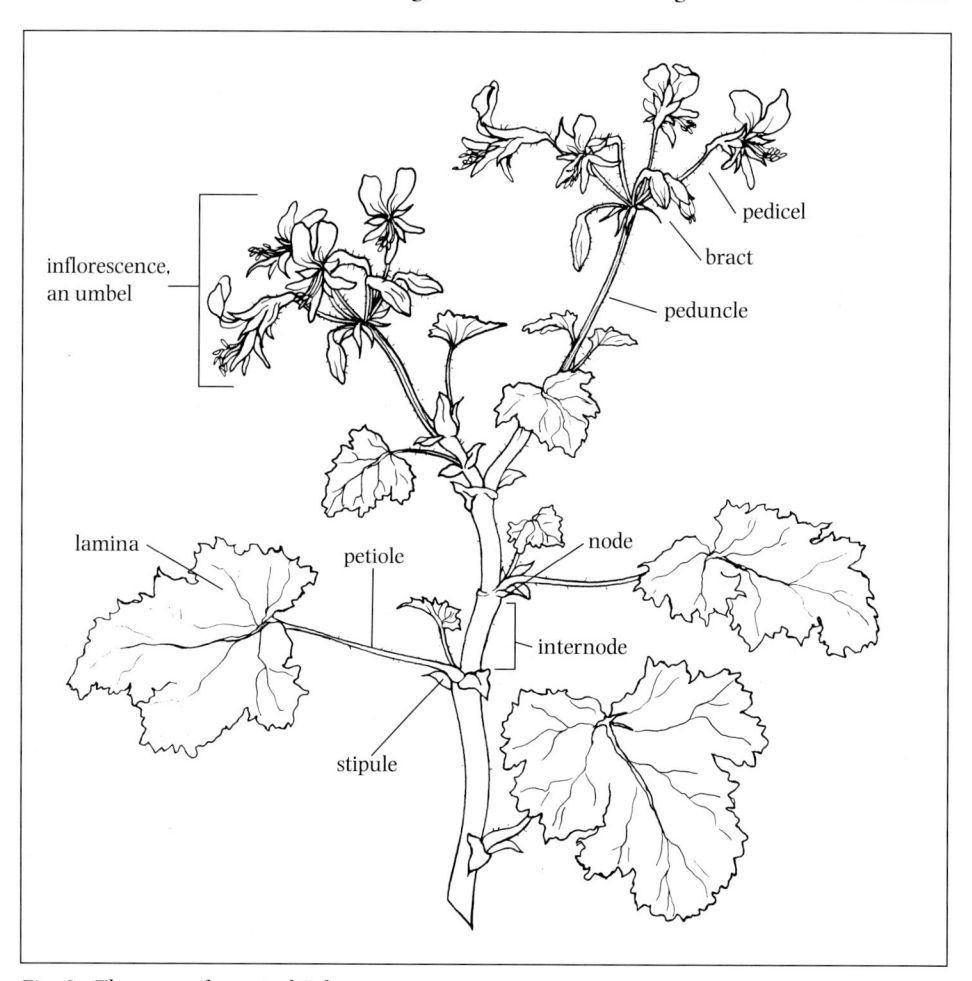

Fig. 6. The parts of a typical *Pelargonium*

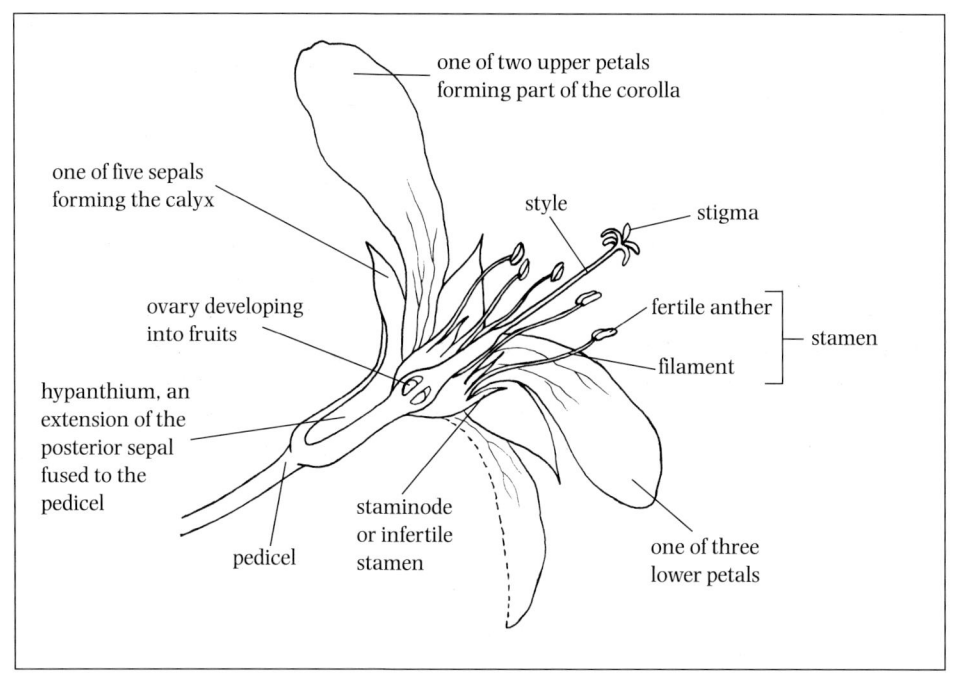

Fig. 7. A section through a typical *Pelargonium* flower

On maturity, the mericarps split apart, and the plumed seed is ejected.

The genus *Pelargonium* is large, with about 230 species arranged into groups of related species. Each of these groups is known as a section. As the species are grown and studied, similarities are easily observed between some, and most of the categorizations into sections appear quite natural. It is however useful in this large genus to understand its classification. If nothing else, it often helps in some cases to judge the ideal cultural conditions, as the species of many sections not only look alike but also grow wild in similar geographical areas and habitats.

The classification of a plant starts when it is first named. As has already been mentioned in the previous chapter, the first pelargonium was what we now know as *P. triste*, but the first name given was a descriptive one as was usual in the seventeenth century, *Geranium Indicum Noctu Odoratum*, which translated means the night-scented geranium from India. This perfectly described the new plant thought to have come from India, related to the geranium and with flowers scented at night. Names at this time were not standardized and several variations may be found in the very early literature. By 1640, Parkinson in his *Theatrum plantarum* calls it *Geranium triste arve Indicum noctu olens*, including the word 'triste' referring to the sad, not bright, colours of the flowers. As more related plants were introduced these descriptive names needed to become longer and became very unwieldy. The African origin of the plants was soon recognized and added to the name as in *Geranium africanum frutescens, folio crasso & glauco, acetosae sapore* used by Commelin in

1703 for *P. acetosum*. This name explains that the plant is a geranium from Africa, which is shrubby with thick glaucous leaves and an acidic sap, a perfectly adequate description. At this time only about a dozen were known so this sort of name was still just about manageable, but imagine what it would have been like after two hundred more species had been discovered.

These early taxonomists were very conscientious about equating any of their long descriptive names with the names used by others. They were also often attached to botanical gardens, gardens of medicinal plants or gardens of noble families and were responsible for the publication of descriptive lists of their acquisitions, some of which are available today. With patience and a great deal of time, it is relatively easy to follow the history of some plants as they moved from garden to garden and across country boundaries.

In 1732, Dillenius suggested that the African species of *Geranium* with unequal and irregular flowers might be called 'Pelargonium' but he did not utilize his own epithet. Six years later Burmann again brought up the name and did employ *Pelargonium* for at least some of the species he depicted.

In 1753 the Swedish botanist, Linnaeus published the first edition of *Species Plantarum* in which his binomial system reduced the names to two words, the generic and specific epithets. He retained the generic name *Geranium* for the twenty *Pelargonium* species as well as what are known today as *Erodium* and *Geranium*, despite previous suggestions to separate the African species. He then chose one significant word from the older descriptive names for the specific epithet. Apart from changing the generic name, the names proposed by Linnaeus are accepted today. The two words which constitute the botanical name of the plant are followed by the authority of the name, that is the author who first published the combination of generic and specific name. These authorities become important when there is confusion in the identity of a plant. This is because the same name may be used by different botanists at different times for more than one distinct species, or several different names applied to the same species. If the authority is cited, the plants and descriptions may be traced and with luck the original herbarium specimen located for comparison. Sometimes an authority is preceded by a previous authors name in brackets indicating that the second author changed part of the name. In most cases, this is because the original description of the species was as a member of the genus *Geranium* and the second author retained the specific name but changed the generic name to *Pelargonium*.

By the latter part of the eighteenth century, the number of new *Pelargonium* species and hybrids had risen dramatically so a means of classifying them became imperative. Cavanilles in 1787 subdivided the species into groups. He separated the African geraniums, which we now know as *Pelargonium*, on the basis of their flower-shape. This group was then subdivided on its leaf characteristics such as the presence of a zone in species such as *P. peltatum*, *P. elongatum* and *P. zonale*, and then on the degree of lobing. Aiton, in 1789, placed species with similar morphology together but did not attempt to formalize the groupings with names. The first serious attempt to do this was by Sweet in 1820, in the first of his five volumes of *Geraniaceae*. He separated many of the more distinct species by creating ten new genera. These differentiations were obviously recognized, but not necessarily considered sufficient to merit the status of unique genera, and de Candolle in 1824 soon reduced Sweet's genera to sections within the original genus *Pelargonium*. Although his classification included many new names for sections which were then subdivided into series, it made the recognition of a plant easier because each grouping was smaller and included far fewer species.

Different classifications were published in the following years such as that by George Don in 1831. George Don was a plant collector for the Royal Horticultural Society's gardens based at Chiswick, but he did not visit South Africa. Dr John Lindley was assistant secretary of the Society at the time and also a botanist, and it was he who first suggested the generic name for the succulent species, *P. carnosum*. Most of the classification systems published were variations or amalgamations of the systems produced by Sweet and de Candolle. Ecklon and Zeyher in 1835 who were prolific plant collectors in South Africa, reinstated the generic status for several sections and created other new ones as well. However, many of the *Pelargonium* species they described were synonyms for existing names, and they seem to have had little influence on classification and the names were soon discarded by Harvey. The Irishman William Harvey became Colonial Treasurer at the Cape in 1835 but combined his duties with his love of plants resulting in the publication in 1860 of *Flora Capensis*.

Reinhard Knuth, later to be appointed Professor of Botany in Berlin, published his work on *Geraniaceae* in 1912 but continued his study of the plants long after. Basically he followed Harvey's naming but included a number of new species. Derek Clifford in 1958 brought much of the information together in a readable format for the general public and it was a remarkable first attempt to bring back interest in the species, after so many years.

Harvey's division of the genus into fifteen sections was followed in principle for over a hundred years until quite recently when the taxonomists in South Africa began a major study of the genus. This investigation is still being carried out by botanists in many parts of the world looking at all aspects of taxonomy. These modern methods cover every conceivable aspect of a plant from its geographical location and habitat to its chromosome size and number, the structure of its pollen coat, the alkaloids and proteins it contains as well as its external morphology and internal anatomy. In many cases, the original classifications have remained very similar to those proposed nearly two hundred years later which is remarkable considering that the early botanists carried out their work relying only on acute observation, with perhaps the aid of a hand lens.

However the recent work has solved many anomalies and is able to show evolutionary trends and relationships. These can rarely be deduced from the external characteristics of a plant as it is not possible from observation alone to decide whether two plants resemble one another because of a close relationship or because of adaptations to similar habitats, climatic conditions or even pollinating insects. Some of the associations are quite obvious and have barely changed, such as the succulent species in the section *Otidia* and the tuberous species of the section *Polyactium* with their rather regular flowers. In other cases, the similar flower structure of *P. praemorsum* and *P. myrrhifolium* with very large upper petals and minute lower petals, encouraged the use of a single category, *Jenkinsonia*, for both. Nowadays, it is clear that, apart from the distinct flower structure, there is little similarity between the two.

The distinctions between some sections have not been considered sufficient to maintain and, for example, the section *Seymouria* with two petalled flowers has been included within *Hoarea*. In other examples, a section has been split as in the case of *Cortusina* where the species found in the more eastern part of its geographical range have been placed in a newly created section *Reniformia*.

In this book, the most recent classification has been followed. The classification used here basically follows that of Professor J.J.A. van der Walt and P.J. Vorster and their team of botanists who have been working on the genus at Stellenbosch University in South Africa for over 20 years. In some sections, detailed works have been carried out more recently and references to any relevant papers are included within the section. Comments about the status of individual species are to be found in the discussion following the species description. The very diverse section *Ligularia* is still being investigated and although this will eventually be subdivided, here, the majority of species have been kept together.

Even with all the battery of modern techniques, there will still be areas of controversy in plant classification. In theory, there should be no argument about plant names but in practice naming is sometimes a matter of opinion and no one is completely right or completely wrong. The perception of the limits of a species may change or new variations may be discovered which alter the limitation of the species. Plants are not static in their evolution and *Pelargonium* appears to be a genus which seems to be continually evolving into new species, although this is not a process which happens in the very short-term. Occasionally a new species may still be discovered in the wild, perhaps even one whose native habitat has been lost, but which has been known in cultivation for hundreds of years and dismissed as a mere garden plant despite its conservation by horticulturists over the centuries. Fortunately, all these factors allow continual study of the genus in all its aspects.

Classification

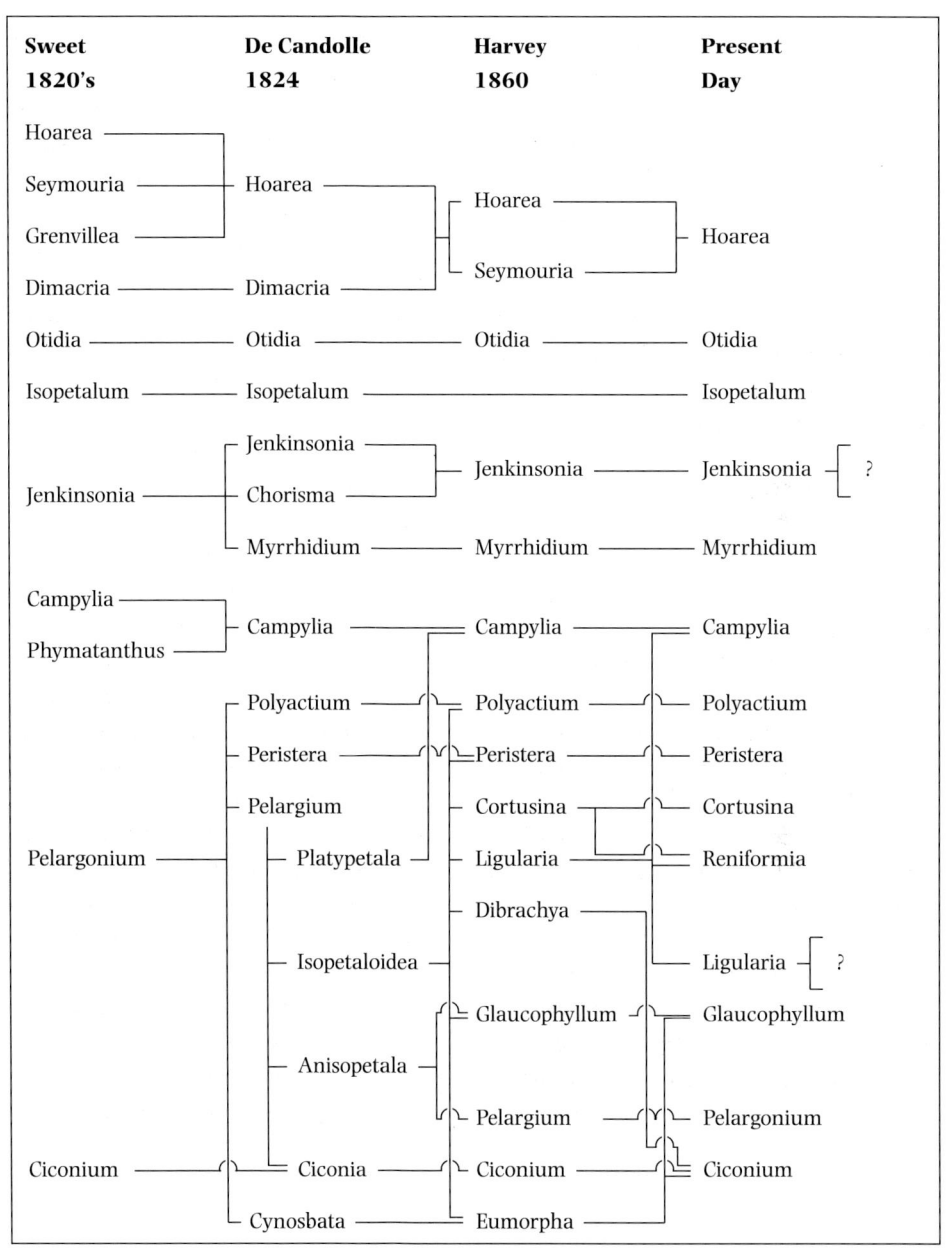

Sweet 1820's	De Candolle 1824	Harvey 1860	Present Day
Hoarea	Hoarea	Hoarea	Hoarea
Seymouria	Dimacria	Seymouria	
Grenvillea			
Dimacria			
Otidia	Otidia	Otidia	Otidia
Isopetalum	Isopetalum		Isopetalum
Jenkinsonia	Jenkinsonia	Jenkinsonia	Jenkinsonia ?
	Chorisma		
	Myrrhidium	Myrrhidium	Myrrhidium
Campylia	Campylia	Campylia	Campylia
Phymatanthus			
Pelargonium	Polyactium	Polyactium	Polyactium
	Peristera	Peristera	Peristera
	Pelargium	Cortusina	Cortusina
	Platypetala	Ligularia	Reniformia
		Dibrachya	
	Isopetaloidea		Ligularia ?
		Glaucophyllum	Glaucophyllum
	Anisopetala	Pelargium	Pelargonium
Ciconium	Ciconia	Ciconium	Ciconium
	Cynosbata	Eumorpha	

Comparative diagram of different methods of classification

3

Introduction to sections

This major portion of the book is designed to describe the majority of species of *Pelargonium*. As far as possible, all those that are in cultivation have been included except for the few that may have been imported by one or two individuals for specialist research. I have grown most and studied live material of all but one or two, and these exceptions have been examined as herbarium specimens.

In their native habitat, the plants often have to struggle for survival in conditions with extreme temperatures and lack of an adequate supply of water for long periods of the year. They are also at the mercy of browsing animals and have to compete for light and space. In contrast, cultivated plants have an easy existence and as a result are often less stunted or contorted, with less woody stems and larger more lush foliage although the flowers are not usually very different in shape, size or colour. Even under different regimes in cultivation, the size and shape of a plant may be variable, so the descriptions and shapes given here are only a guide to the relative proportions of different parts of the plant. Photocopies of leaves will help in the identification of the plants. These should be used in association with the description as the leave shape may be variable even in cultivation. The majority are slightly reduced in size.

A key has been developed for each section which should enable the reader to track down and name an unknown species within a section. These keys to the sections have been prepared from cultivated examples of the species although it should be possible to use them also on wild collected material. They are designed to be capable of use by anyone with an interest in the genus and not just for the experienced botanist. The botanical terms have therefore been kept to a minimum and although a hand lens might be useful in some cases, there is no necessity for any other equipment.

However the sections themselves are less easy to define using only readily visible morphological characteristics. In recent taxonomic studies, many of the sections have been more accurately redefined using modern botanical techniques such as chromosome counts or pollen analysis, but these methods are totally impractical for the majority of *Pelargonium* enthusiasts. From the gardener's point of view, there are numerous exceptions amongst the plants within each section which cause obstacles in the process of classification. To facilitate this procedure, brief generalized descriptions of each section are given here and a table has been created which should indicate the main characteristics of each. A much more detailed description will be found under the relevant section heading. After growing and handling the plants, as well as observing them in detail, the section to which any one pelargonium belongs may be instinctively recognized by a combination of characteristics which can be more difficult to define accurately in words. For an unidentified species the table should help to indicate in which section the plant is classified, but if the relevant key does not result in an appropriate identification another should be attempted.

For those who have never used a key for the identification of an unknown plant, a few hints may help. A botanical key is a means of identifying an unknown plant without having to read through every description which, with large numbers, becomes very tedious. A key is basically a series of questions by which many species can be reduced into smaller and smaller units until eventually there is only one possible choice remaining, which should be the name of the plant in question. For any one group of plants, any number of keys may be created using a range of different characteristics. It is important to follow the key carefully, reading all options, and then the description of the resulting plant. If the description does not agree with the features of the plant in question, it is worth trying again. Do remember however, that plants are variable even within one species in cultivation and do not always fit neatly under one name. This is especially true of plants grown from wild collected seed rather than vegetatively propagated plants and the final identification has to be the result of all characteristics. The following example is a key for four completely hypothetical plants: a shrub with white flowers; an annual climbing plant with white flowers; a red-flowered plant with leaves divided into three leaflets; and a yellow-flowered plant with undivided leaves.

1 Flowers white   2
   Flowers not white   3
2 Plant a shrub   'A'
   Plant an annual climber   'B'
3 Flowers red; leaves divided into
   3 leaflets   'C'
   Flowers yellow; leaves not divided   'D'

In this example, the first question to ask is 'are the flowers white?' If the answer is 'yes', then go to the options under number 2. If the answer is that the flowers are not white, then go to question 3 and so on. The four plants are quickly identified:

'A' is the shrub with white flowers;
'B' is the annual climbing plant with white flowers;
'C' is the red-flowered plant with leaves divided into three leaflets;
'D' is the yellow-flowered plant with undivided leaves.

If at question 3, the flowers were blue or the flowers red but the leaves never divided, it would indicate that the key has not included the unknown plant in question or perhaps that the plant was a cultivated selection or a hybrid. This is a very simplified example but should serve to explain how a key works to those who might be uneasy about using one for the first time.

Table showing the main features of each section

Key: xxx all or the majority of species in the section
      o   no species in the section
      x   some species in the section

This table is a guide to indicate the characteristics of each section but there are often one or two species in a section which do

Table showing the main features of each section

Section Feature	Campylia	Ciconium	Cortusina	Glauco- phyllum	Hoarea	Isopeta- lum
Tubers present	o	o	x	o	xxx	o
Swollen or succulent stems	o	x	xxx	o	o	xxx
Stems very short	xxx	x	x	o	xxx	o
Stems spiny	o	o	x	o	o	o
Stipules, or petioles persistent	xxx		xxx	x	xxx	xxx
Perennials dying down for part of year	o	o	x	o	xxx	o
Annual or short-lived perennial	o	x	o	o	o	o
Deciduous		o	xxx			xxx
Evergreen		xxx	o			o
Leaves aromatic	o		x	o		o
Glaucous leaves	x	x	x	xxx	x	o
Flowers scented	o	o	o	o		o
Flowers very irregular	x	x	o	xxx	x	o
Flowers almost regular	xxx	xxx	xxx	o	o	xxx
Flowers with 2 petals	o	o	o	o	x	o
Upper 2 petals very much larger	x	o	o	o	x	o
Hypanthium shorter than pedicel	xxx		o	o		xxx
Hypanthium longer than pedicel	o	xxx	xxx	x	xxx	o
Fertile stamens	5 or 7	5 or 7	6 or 7	5–7	2–5	5
Inflorescence branched			o	o	x	xxx

Jenkin-sonia	Ligularia	Myrrhi-dium	Otidia	Pelar-gonium	Peristera	Poly-actium	Reni-formia
x	x	o	x	o	x	xxx	
x	x	o	xxx	o	x	x	o
x	x	o	o	o	o	x	x
x	x	o	x	o	o	o	o
x	x	o	x	o	x	o	xxx
x	x	x	o	o		xxx	o
o	o	xxx	o	o	xxx	o	o
x		xxx					o
o			o	xxx			xxx
x	x	o	o	xxx	x	o	xxx
o		x	x	x	o	x	x
x	o	o	o	o	o	xxx	o
xxx	xxx	xxx		x	o	o	
o	x	o			x	xxx	
x	o	o	o	o	o	o	o
xxx		xxx	o	x	o	o	o
o			xxx	xxx	xxx	o	o
xxx			o	o	o	xxx	xxx
7	5 or 7	5 or 7	5	7	4–7	6 or 7	7
o	x	o	x	x			xxx

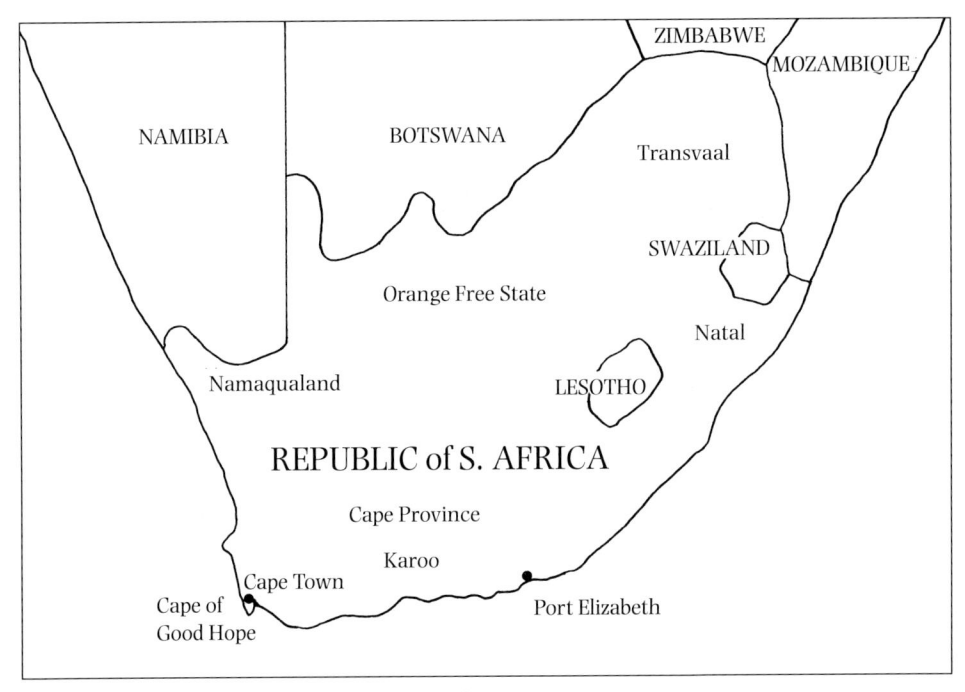

Fig. 8. The geographical areas of southern Africa

not conform in all characteristics. To determine the section for an unknown species, try the one which conforms in the majority of details. Where a box of the table has not been completed, the characteristic in question is variable within the section and of little diagnostic value. A map of the geographical regions of southern Africa may also help to locate the natural distribution of the species (fig. 8).

Synopsis of sections

Campylia
Low-growing plants with tufted or rosette habit. Flowers with short hypanthium.

Ciconium
Large plants often with fleshy stems and simple palmately-veined leaves. The flowers have a long hypanthium and the petals tend to be similar in size, or the lower three slightly larger than the upper two.

Cortusina
Plants with thick semi-succulent stems covered with persistent stipules or petioles. Flower with similar petals and a long hypanthium.

Glaucophyllum
Shrubs or subshrubs often with leathery glaucous leaves. Flower very irregular with long hypanthium.

Hoarea
Stemless plants with tubers. Leaves usually dying down as flowers appear.

Isopetalum
Succulent plant with more or less regular white flowers and undivided leaves.

Jenkinsonia
Plants with variable habit and palmate leaves. Upper petals always very much larger than lower which may even be absent.

Ligularia
This diverse section contains plants with almost any combination of characters.

Myrrhidium
Herbaceous, often short-lived plants with pinnately-divided leaves. Upper petals always very much larger than lower.

Otidia
Succulent plants with irregular, usually white flowers and incised leaves. Fertile stamens five.

Pelargonium
Usually quite large shrubs or subshrubs many with aromatic foliage. Flowers with upper petals larger than lower and seven fertile stamens.

Peristera
Annuals or short-lived perennials. Flower small with short hypanthium.

Polyactium
Plants with tubers. Leaves usually present at same time as flowers. Flower often night-scented and more or less regular in shape with long hypanthium.

Reniformia
Stems with persistent stipules or petioles. Leaves simple with long petioles. Flower irregular with long hypanthium and seven fertile stamens.

5

The *Pelargonium* species

Campylia (Sweet) de Candolle

The first use of the name for this section was by Sweet when he separated *P. ovale* into a new genus, *Campylia*, in 1820. This name was taken from the Greek Camplyos, meaning curved, on account of the two bent staminodes found in this and some other species. Others of the present section such as *P. tricolor*, with erect staminodes, he assigned to yet another genus, *Phymatanthus*, meaning wart-flower because of the shiny raised bumps at the base of the two upper petals. These two genera were amalgamated into the one section *Campylia* by de Candolle a few years later, and the most recent work in South Africa has added the newly described species *P. ocellatum* as well as some from other sections.[1]

Campylia includes some of the most attractive plants of the genus but also some of the most difficult to grow. The plants are characteristically small and tufted with short stems, often covered with the remains of the petiole bases. Some have a rosette habit and all have an extensive root system. The leaves are not usually deeply divided but have long petioles with conspicuous membranous stipules at the base. The flowers are often rather open or 'flat' in shape with 5, occasionally 4, often rather broad petals, a long pedicel and 5 or 7 fertile stamens. The hypanthium is usually very short and in some cases the only sign is a slight depression at the base of the upper sepal. In some species such as *P. ovale*, the upper petals have constricted or auricled bases whereas *P. tricolor* has raised dark glossy warty spots at the base of the upper petals, and when these become wet the colour tends to run onto the rest of the petals. These patches are false nectaries and by attracting insects are important in the pollination of the flowers. *P. capillare*, *P. incarnatum* and *P. ocellatum* have similar shiny patches but these are less noticeable than in *P. tricolor*.[2]

In the wild, this section is centred in south-western and southern parts of the Cape Province where the plants tend to grow in mountainous regions with very freely-draining soils. Their extensive root system enables the plants to tap water from well below the surface and the rosette habit helps to reduce loss of water by transpiration in the hot dry summers and the intense light.

Most members of this section had been discovered and were cultivated by the early part of the nineteenth century. For example, Jacquin describing the plants in the gardens of the Emperor Joseph II at Schönbrunn in Austria, in 1797, illustrates and names several forms of *P. ovale* being cultivated at the time, and some thirty years later Sweet was illustrating several more species. Many others with attractive and unusual flowers were obviously garden hybrids, perhaps attempts by the enthusiasts of the day to produce plants with less exacting cultural requirements. The only hybrid commonly found today, which has been grown under a variety of names, is *P.* 'Splendide' and most plants sold in the past as *P. tricolor* appear to be this cultivar.

Species of this section are being used today for hybridizing, especially in Germany, but few, if any of the resulting plants are widely available, perhaps because of the difficulty of propagating, transplanting and growing them successfully.

Although cuttings of non-flowering shoots root quite easily, great care should be taken when transplanting as the plants resent any root disturbance. Watering is critical and the compost should be damp but never wet. Hormone rooting powder and a fungicide may be found to improve the success rate. Some species appear to be self-sterile but when produced, the seed usually germinates as readily as that of species of other sections although the young plants are not always easy to maintain. They certainly need bright but not direct sunlight in the hottest months. As they naturally have extensive root systems perhaps one answer is to grow them in larger containers so that the roots may remain cool. However, the most important factor is an exceedingly free-draining compost. A mix of about half horticultural silver sand to half peat or soilless compost with the regular application of a liquid feed during the growing season seems to work. If plants become old and lose their vigour, they may be cut back to allow new shoots to regenerate the plant.

The basic chromosome number is 10 and most have 20 or 40 small chromosomes; there appears to be some affinity between this section and the sections *Pelargonium* and *Glaucophyllum*. Although the habit of most of the species of the two sections is remarkably different and the basic chromosome is not the same, the flower structure of some species such as *P. elegans* and *P. betulinum* shows great similarity and a few garden hybrids between the two have been raised.

1	Flowers with a clearly visible shiny embossed spot at the base of the two upper petals	2
	Flowers without an embossed spot on the upper petals	4
2	Hypanthium over 10 mm long	**capillare**
	Hypanthium under 10 mm long, usually much less	3
3	Leaf blade pinnately veined and toothed	**tricolor**
	Leaf blade trilobed or trifoliate	**ocellatum**
4	Upper and lower petals dissimilar in size, shape or colour	5
	Upper and lower petals similar in size, shape and colour	**incarnatum**
5	Petals usually 4	**caespitosum**
	Petals 5	6
6	Fertile stamens 5	7
	Fertile stamens 7	9
7	Petiole longer than leaf blade	**ovale**
	Petiole shorter than leaf blade	8
8	Upper petals constricted at base; pedicel over 2 cm ($4/5$ in) long	**coronopifolium**
	Upper petals not constricted at base; pedicel under 5 mm long	**oenothera**
9	Flowers to over 4 cm (1 $3/5$ in) across; leaves with rounded apex	**elegans**
	Flowers under 3 cm (1 $1/5$ in) across; leaves with acute apex	**setulosum**

P. caespitosum Turczaninow, *Bulletin de la société des naturalistes de Moscou*, **31**: 420 (1858)

MEANING: tufted, referring to habit
SECTION: **Campylia**
ILLUSTRATION: *Pelargoniums of southern Africa*, **2**: 17.

A small tufted plant to about 15 cm (6 in) tall, woody at base. Leaves aromatic, elliptic, often folded towards midrib, irregularly toothed, grey-green, hairy, c. 40 × 15 mm; petiole exceeding leaf blade, base semi-persistent; stipules c. 1 cm ($2/5$ in), narrow. Flowers deep-pink or purplish-pink, on long peduncle to 10 cm (4 in); upper petals with darker markings, spathulate, curled inwards at base to form a tubular base, reflexed, c. 10 × 5 mm; lower petals smaller, unmarked; hypanthium a small cavity; pedicel 2–3 cm ($4/5$–1 $1/5$ in); fertile stamens 5, unequal.

RECOGNITION: small tufted plant with 4 hairy petals.

This species grows in high mountains in south-western Cape Province and was first collected in 1838 by Drege but not fully recognized until much later. It is rarely grown in collections in Europe and is as difficult to maintain as any of the section. A second subspecies subsp. *concavum* Hugo, has recently been distinguished, differing from the type by the hairless petals and narrow, greener leaves.[3]

P. capillare (Cavanilles) Willdenow, *Species plantarum*, **3**: 660 (1800)

MEANING: thread-like, perhaps referring to long, thin petioles

SECTION: **Campylia**

ILLUSTRATION: *Pelargoniums of southern Africa*, **2**: 24.

This species is somewhat similar to *P. tricolor* but easily distinguished from it. The flowers have pinkish-red petals, darker on the underside, the upper with raised embossed patches and all with darker marks. Seven fertile stamens are present and the conspicuous hypanthium is often over 1 cm ($^2/_5$ in) long. The leaves have very long thin petioles, 2–6 cm ($^4/_5$–2 $^2/_5$ in) long, up to twice the length of the lamina which is more deeply and irregularly divided than that of *P. tricolor*. Like others of the section, it grows wild in the mountains of south-western Cape Province. Nowadays it is rarely found in collections in Europe even though it was first illustrated over 200 years ago by the Spanish botanist, Cavanilles.

P. coronopifolium Jacquin, *Icones plantarum rariorum*, **3**: 9, t. 526 (1792)

MEANING: leaves resembling *Coronopus*, a species found wild in Europe

SECTION: **Campylia**

ILLUSTRATION: *Pelargoniums of southern Africa*, **2**: 35.

A small upright plant, which can form colonies as a result of the branching underground system, the older woody stem base covered with remains of persistent petioles. Leaves somewhat rough, linear to narrow elliptic, irregularly toothed along the upper half, the sides of the lamina often folded inwards, *c.* 7 x 0.5 cm (2 $^4/_5$ x $^1/_5$ in); petiole 2–4 cm (4/5-1 3/5 in); stipules narrow to 1 cm ($^2/_5$ in). Flowers pink to purple, occasionally white; peduncle to 5 cm (2 in) usually bearing 2–3 flowers; upper petals with some darker markings, obovate, *c.* 15 x 5 mm, auricled at base; lower petals slightly smaller, unmarked; hypanthium well developed to over 5 mm; pedicel to 4 cm (1 $^3/_5$ in); fertile stamens 5.

RECOGNITION: could be confused with *P. oenothera* which is distinguished by the auricled upper petals or with *P. ovale* subsp. *veronicifolium* which has petioles much longer than the lamina.

Some forms of this species have such narrow leaves that they have been considered by some to merit specific or subspecific status such as *P. gramineum* Bolus (1889), *P. angustissimum* Knuth (1912) and var. *lineare* Harvey (1860). The species is not widely cultivated and for this publication all have been included under the one specific name. However, if the different morphological forms become more widely grown in gardens, distinct cultivar names may become appropriate. It grows wild in mountainous regions of south-western Cape Province but may also be found at lower altitudes. F. Masson collected it for Kew in 1791 and it was described by the Austrian botanist, Jacquin 3 years later.

P. elegans (Andrews) Willdenow (1800), *Species plantarum*, **3, 1**: 655 (1800)

MEANING: elegant

SECTION: **Campylia**

ILLUSTRATION: Sweet, **1**: 36; *Pelargoniums of southern Africa*, **1**: 14.

A small erect tufted plant. Leaves green, hairy or glabrous, broad ovate to almost orbicular, with truncate to cordate base, margin often reddish with rather coarse, sharp teeth, to 4 x 3 cm (1 $^3/_5$ x 1 $^1/_5$ in) but often less; petioles to 10 cm (4 in); stipules triangular, to 15 mm. Flowers large, pale to deep-pink; peduncle usually 3–4 flowered, to 10 cm (4 in); upper petals broadly obovate, with darker feathering, *c.* 25 x 14 mm; lower petals unmarked, narrower; hypanthium *c.* 1 cm ($^2/_5$ in); pedicel to 4 cm (1 $^3/_5$ in); fertile stamens 7.

RECOGNITION: plants, especially those from the more westerly areas, may be confused with *P. ovale*, but *P. elegans* has less hairy, almost leathery, green leaves with the lamina tending to be held upright on a rigid petiole.

Whereas many species of *Pelargonium* have a rather limited distribution in the wild which is usually related to rainfall, *P. elegans* is found in two geographically separated areas in eastern and south-western Cape Province often near the coast. It was introduced by the British nursery firm of Lee and Kennedy in 1795 and forms a very attractive small plant with large flowers produced over a long period from early spring to early summer.

P. incarnatum (L'Héritier) Moench, *Supplementa ad methodum plantas*, 295 (1802)

MEANING: referring to flesh-coloured flowers
SECTION: **Campylia**
ILLUSTRATION: *Botanical magazine*, **8**: 261; Sweet, **1**: 94.

A spreading plant, to 20 cm (8 in) high, woody at the base with reddish stems. Leaves crowded, 3-lobed, the lobes irregularly toothed and the terminal lobe often pinnately lobed, 15–20 mm long; petiole reddish, to 10 cm (4 in), base persistent; stipules partly fused to petiole. Flowers regular, basically pale pink, dark-red at base, and with concentric zones of deep-pink and white, 4–5 on peduncle to over 5 cm (2 in); petals obovate, 10 × 7 mm; hypanthium represented by a shallow cavity at the base of the upper sepal; pedicel 1–4 cm ($2/5$-1 $3/5$ in); fertile stamens 5 with very short filaments.

RECOGNITION: Similar to *P. setulosum* in habit and foliage but distinguished by the regular, pale salmon pink flower.

First sent to Kew by F. Masson in 1787, collected from the mountains of south-western Cape Province, this species was also raised from seed collected in South Africa for Colvill's nursery in Chelsea. For many years it was considered to be a member of the genus *Erodium* but despite the morphological similarity, for several other botanical reasons as well as the discovery of a hybrid with the *Pelargonium* species *P. patulum* and the fact that it was the only native *Erodium* found in the southern hemisphere, it is far more reasonable to include it within *Pelargonium*.[4]

P. ocellatum J.J.A. van der Walt, *South African journal of botany*, **56**: 467 (1990)

MEANING: refers to dark spots on upper petals
SECTION: **Campylia**
ILLUSTRATION: *South African journal of botany*, **56**: 468 1990.

This recently recognized species grows in semi-arid regions of south-west Cape Province and may be distinguished from *P. tricolor* by the trifoliate or 3-lobed leaves to 3 × 2.5 cm (1 $1/5$ × 1 in) with toothed linear segments and from *P. capillare* by the 5 fertile stamens. It forms a small tufted plant, the stem covered with the remains of the petiole bases. The branches of the inflorescence bear up to 3 white to pale pink flowers each to nearly 2 cm ($4/5$ in) across with dark-red embossed spots at the base of the upper obovate petals. The hypanthium and pedicel each reach to 10 mm.

P. oenothera (Linnaeus filius) Jacquin, *Icones plantarum rariorum*, **3**: 9 t. 525 (1792)

MEANING: leaves like a species of *Oenothera*
SECTION: **Campylia**
ILLUSTRATION: *Pelargoniums of southern Africa*, **2**: 101.

Somewhat similar to *P. coronopifolium*, this is a plant of more botanical than horticultural interest and is only cultivated by a few enthusiasts of the genus. It has attractive grey-green, softly-hairy leaves, usually broader than those of *P. coronopifolium*, never folded but toothed along the whole margin. The slightly smaller and paler flowers are formed in a tighter inflorescence caused by the much shorter hypanthium and pedicel and the upper petals are not auricled at the base. Two of the staminodes are somewhat curved, a feature seen more clearly in *P. ovale*. It grows in a similar region but often at lower altitudes and was known in Europe well before the end of the eighteenth century.

P. ovale (Burman filius) L'Héritier, Aiton, *Hortus Kewensis*, **2**: 429 (1789)

MEANING: oval, referring to leaf shape
SECTION: **Campylia**
ILLUSTRATION: Sweet, **1**: 88; *Pelargoniums of southern Africa*, **1**: 31.

A low-growing plant, the woody base covered with persistent stipules. Leaves ranging from almost rounded to lanceolate, hairy, often grey-green, 20–40 x 5–30 mm; petiole exceeding leaf blade; stipules triangular to 15 mm. Flowers white to deep-pink, peduncle to 10 cm (4 in) bearing about 5 flowers; upper petals with darker markings, obovate, to 25 x 15 mm, with auricles; lower petals narrower, unmarked; hypanthium 1–10 mm; pedicel to 5 cm (2 in); fertile stamens 5, with two of the remaining staminodes curved.

RECOGNITION: Some forms of the species might be confused with *P. elegans* but the leaves are usually much more hairy and in the latter, the petals are not auricled and the flowers have seven fertile stamens.

One of the easier species of the section to cultivate, this species makes a very attractive plant with grey-green leaves and relatively large soft pink flowers. The first of the section to be described, it has been grown since its early introduction in the middle of the eighteenth century. Francis Masson collected slightly different plants for Kew on at least three of his expeditions in 1774, 1790 and 1794, and all the variations at that time had already been given different specific names based mainly on their foliage characteristics. Several garden hybrids were raised in the early nineteenth century and described by Sweet although perhaps the only one commonly seen today is 'Splendide' (see under *P. tricolor*). In the wild it is found in mountainous regions from the winter rainfall areas of south-western to southern and eastern Cape Province where there is some rain at all times of the year.

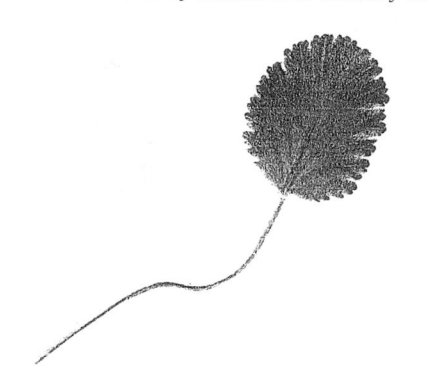

Despite the variability in the wild and the numerous names under which it has been known in the past, only three subspecies are now recognized.[3] The one usually found in cultivation is *P. ovale* subsp. *ovale*, with ovate leaves, the upper and lower petals of similar length and a hypanthium over 2 mm long. *P. ovale* subsp. *veronicifolium* (Eckl. & Zeyh.) Hugo, was first differentiated as a species by Ecklon and Zeyher in 1835 and may be distinguished by the very narrow leaves and tiny lower petals less than half the length of the upper so that the stamens protrude conspicuously beyond them. This subspecies is found in the eastern end of the species range. In *P. ovale* subsp. *hyalinum* Hugo, found at high altitudes in a restricted area of the south-western region of Cape Province, the flowers are dark-pink, the petals may be hairy, the lower petals quite broad, the hypanthium minute and the broadly ovate leaves densely covered with long hairs. *(Plates 1 & 3)*

P. setulosum Turczaninow, *Bulletin de la société des naturalistes de Moscou*, **31**: 422 (1858)

MEANING: referring to stiff hairs on leaf
SECTION: **Campylia**
ILLUSTRATION: *Pelargoniums of southern Africa*, **2**: 126.

For many years this species was included in the section *Ligularia*. However, the habit, leaf and flower shape clearly resemble *P. elegans* although the flowers are smaller and the leaves more leathery, similar to those of *P. incarnatum*. It grows in the more mountainous regions of south-western Cape Province. The flowers are white or pale pink blotched on the upper petals with dark-red, to about 2.5 cm (1 in) across. The leaves usually have a cordate base, acute apex and the stipules to 1 cm ($^2/_5$ in), are narrow triangular. It was first named from plants collected by Zeyher in 1857 under his collection number 2084.

P. tricolor Curtis, *Botanical magazine*, **7**: 240 (1793)

MEANING: 3-coloured, referring to flower
SECTION: **Campylia**
ILLUSTRATION: Sweet **1**: 43; *Pelargoniums of southern Africa*, **1**: 48.

A small plant to about 30 cm (12 in) high, the basal woody part of the stem covered with the remains of stipules. Leaves grey-green, hairy, narrow elliptic with sharp irregular teeth, often with two larger lobes at the base, usually *c.* 3 × 1 cm (1 $^1/_5$ × $^2/_5$ in); petiole about same length as leaf blade; stipules narrow triangular. Flowers almost regular in shape; peduncle 2–4 flowered, to 4 cm (1 $^3/_5$ in); upper petals suborbicular, deep-red with a glossy, raised black spot at base, *c.* 15 × 12 mm; lower petals equal in size or slightly larger, white, unspotted; hypanthium 1–2 mm; pedicel 1–2 cm ($^2/_5$–$^4/_5$ in); fertile stamens 5.

RECOGNITION: flowers normally resembling a wild pansy, the upper petals deep red with a black raised spot and the lower petals white. Plants are occasionally found with white or mauve-pink flowers still bearing the dark patches at the base of the upper petals.

P. tricolor is a species found growing wild, often under other shrubs, in dry sandy soils in the foothills of the mountains in a relatively small area of south-western Cape Province. Plants growing in the shade have larger flowers which are redder in colour while those growing in full sun are smaller with deeper more purple-red upper petals. It was collected by Francis Masson on his expedition of 1791, flowered in cultivation in 1792 and named and illustrated in the *Botanical Magazine* two years later. By propagating non-flowering shoots in late autumn, using a compost of at least half grit, and being careful never to overwater, it can be grown successfully, but the true species is rare in cultivation and for many years has been confused with the cultivar 'Splendide'.

This cultivar is easier to propagate and maintain but has been incorrectly known by a variety of names including *P. tricolor*, *P. tricolor arborea*, *P. violareum* and *P. violaceum*. In 1953 it received a Preliminary Commendation from the Royal Horticultural Society as *P. tricolor arborea* when exhibited by Lord Astor of Hever Castle and in 1964 an Award of Merit. Following correspondence with the Royal Botanic Gardens at Kew who noted that it appeared to be a putative hybrid between *P. tricolor* and *P. ovale* subsp. *ovale*, it was named 'Splendide', in 1958, the name under which it should be known today. It may be distinguished from *P. tricolor* by the long petioled, grey-green, hairy leaves which are broadly ovate and more regularly serrated, and the slightly larger flowers and dark purple stamens. Seedlings from this cultivar have been raised showing a range of characteristics from both parents.

There are several illustrations of similar, but not identical, plants under other names in the literature of the early nineteenth century such as *P. elatus* in Sweet **1**: 96. There may be other hybrids in cultivation, but new cultivar names should not be applied unless the differences are very clearly defined and published to avoid even more confusion in the future. One named 'Renate Parsley' has recently been named which is a result of a cross between *P. tricolor* and a very deep purplish-pink form of *P. ovale*. This is quite distinct with pale pink lower petals, each petal having a slighty darker pink marking forming a ring around the flower. Other hybrids have been raised more recently such as 'Islington Peppermint', with the peppermint-scented leaves of *P. tomentosum* and tiny tricoloured flowers.
(Plates 2 & 3)

P. 'Splendide'

Ciconium (Sweet) Harvey

For horticulture, this is an extremely important section as it includes several species involved in the ancestry of both the zonal and ivy-leaved pelargoniums. Older classifications were based on the habit of these species. Some which are tall and erect, or scandent plants with thick fleshy stems becoming woody as they age, like *P. inquinans*, were included by Sweet in a distinct genus, *Ciconium*. Those with the trailing habit of *P. peltatum* were in a second distinct genus, *Dibrachya*. Ecklon and Zeyher in 1835, proposed a third genus, *Eumorpha* for the smaller, herbaceous species such as *P. alchemilloides* which in general, tend initially to form neat, compact plants with short stems but then become rather untidy as trailing, leafy branches develop bearing the flowers. The three groups are still morphologically distinct but modern botanical taxonomic methods indicate that there are sufficient similarities to unite them. There are also three species, *P. aridum*, *P. articulatum*, and *P. barklyi* in the present section *Ligularia* which are close in their morphology to members of this section with a similar chromosome number and which could easily be included here.

If not pruned back for the winter, nearly all are evergreen in cultivation, with simple, palmately lobed, but not repeatedly divided, leaves. The flowers are generally rather regular in shape with five more or less equal petals which tend to be in shades of red and pink or white but less frequently in the purplish shades associated with species in other sections. Unusually the lower petals are often equal in size or sometimes even larger than the upper. They also tend to have fewer of the lines and spots on the upper petals associated with most other pelargoniums. There are seven, sometimes fewer, fertile stamens of which two or more may have extremely short filaments or all filaments may be united for almost their whole length. The hypanthium is long, mostly over five times the pedicel length. Typically the bud is noticeably bent downwards but the pedicel straightens as the flower opens then reflexes again as the fruit matures. The basic chromosome number is $x = 9$ and the majority have 18 chromosomes. *P. caylae* has twice this number, *P. elongatum* only 8 and in the variable and complex *P. alchemilloides* group several different counts have been found.

The species grow wild in the eastern part of Cape Province in the area of summer rainfall but *P. zonale* extends westwards while others may be found further north as far as East Africa and Yemen. Some are found in areas of higher humidity than is usually expected for the genus and some in grassland or almost semi-tropical locations, although in the regions nearer the equator they grow at much higher altitudes and therefore avoid the truly hot, humid tropical conditions. None grow in the very arid inland areas and therefore need fewer adaptations to harsh environmental conditions and limited water supplies. *P. zonale* and *P. inquinans* are the major parents of the zonal cultivars. Even though both are among the earliest introductions of the genus at the beginning of the eighteenth century, it is surprising how few hybrids were described until over a hundred years later when the main development of the zonals began.

Members of this section are as easy to grow as any zonal cultivar but a few from the more northerly parts of South Africa and from East Africa may be slightly more tender and *P. caylae* from Madagascar is shy to flower. Many come from areas of higher rainfall or sometimes even from climates with hot humid summers. Watering is generally less critical for these species and a very free-draining compost less vital. The elongated stems of some species may be cut back after flowering; the shrubby and scandent plants benefit from judicious pruning to maintain a bushy habit if the aim is to produce an attractive looking plant. Propagation from cuttings is straightforward and seed normally germinates easily though not necessarily quickly.

1	Plants trailing; leaves ivy-shaped	2
	Plants not with this habit	4
2	Leaves over 3 cm (1 $^{1}/_{5}$ in) long	3
	Leaves under 2 cm ($^{4}/_{5}$ in) long	**'Saxifragoides'**

3	Leaves peltate	**peltatum**
	Leaves not peltate	**'Lateripes'**
4	Petals <1.5 times as long as wide	5
	Petals >1.5 times as long as wide	9
5	Leaves without a zone	6
	Leaves with a zone	8
6	Flowers pink	**acreaum**
	Flowers red	
7	Leaves rounded	**inquinans**
	Leaves angular	**tongaense**
8	Flowers white	**multibracteatum**
	Flowers salmon-pink	**frutetorum**
9	Stamens 5 (sometimes 4 or 6)	10
	Stamens 7	13
10	Petals narrower than sepals	**stenopetalum**
	Petals broader than sepals	11
11	Leaves glaucous	12
	Leaves green	**caylae**
12	Leaves cuneate, petals linear	**acetosum**
	Leaves truncate or shallow-cordate; petals narrow obovate	**salmoneum**
13	Upper petals > 15 mm long	14
	Upper petals < 15 mm long	15
14	Leaves conspicuously hairy	**transvaalense**
	Leaves not usually hairy	**zonale**
15	Flowers grey-yellowish-green in colour	**quinquelobatum**
	Flowers cream, white or pink	16
16	Stipules narrow, linear	**elongatum**
	Stipules broad	17
17	Upper petals exceeding lower; stamens often curved upwards	**mutans**
	Upper and lower petals more or less equal; stamens straight	**alchemilloides**

P. acetosum (Linnaeus) L'Héritier, Aiton, *Hortus Kewensis*, **2:** 430 (1789)

MEANING: acid referring to acid sap

SECTION: **Ciconium**

ILLUSTRATION: *Botanical magazine*, **3:** 103; *Flowering plants of Africa*, **45:** 1800; *Pelargoniums of southern Africa*, **1:** 2.

Branching, subshrub with brittle stems, hairless except for glandular hairs on calyx and hypanthium, to *c.* 60 cm (24 in) in height in cultivation. Leaves, fleshy, glaucous, often almost cupped, sometimes with reddish crenate margin, broad obovate, cuneate base, 2–6 × 1–2 cm (⁴/₅–2 ³/₅ × ²/₅–⁴/₅ in); petiole *c.* 1 cm (²/₅ in); stipules triangular. Flowers pale to salmon-pink, very irregular; inflorescence 2–7 flowered; peduncle to 10 cm (4 in); upper petals very narrow, erect, marked with darker veins, 25 × 5 mm; lower petals similar but unmarked, slightly broader and spreading; hypanthium 25–30 mm; pedicel *c.* 5 mm; fertile stamens 5.

RECOGNITION: blue-green succulent leaves and pale pink narrow-petalled flowers.

Casper Commelin illustrated *P. acetosum* in the list of plants known in the medicinal plant garden in Amsterdam in 1703 and it was growing at Chelsea by 1724. Perhaps it was thought that the sorrel tasting leaves had some curative properties although they have no known medicinal value. There is no record of how it came to be in the garden but it does grow in the eastern Cape Province, near where ships could have stopped for water on their voyages to and from the Indies at that time. It needs drier conditions than many others of the section and is an attractive plant flowering over a long period. No obvious hybrids have resulted from this species and the narrow petals were probably disregarded during the development of the zonals as the hybridizers of the time were looking for large, broad-petalled flowers. However, accidental crosses could have produced cultivars which may have been involved in the early ancestry of the modern zonal group. A form with cream-edged leaves may occasionally be obtained and one with scarlet flowers is mentioned by Andrews although he comments that he had not been able to see it himself.

P. acraeum R.A. Dyer, *Flowering plants of South Africa*, **20**: 779 (1940)

MEANING: dweller on the heights because it grows in mountainous regions

SECTION: **Ciconium**

ILLUSTRATION: *Flowering plants of South Africa*, **78**: 779; *Pelargoniums of southern Africa*, **2**: 1.

Rather a spreading plant, stems thick, fleshy barely 50 cm (20 in) tall unless allowed to scramble through tall plants,. Leaves almost orbicular, with short hairs, unzoned, lobed and coarsely, irregularly dentate, 8–10 cm (3 $^1/_5$–4 in) across; petiole 3–5 cm (1 $^1/_5$-2 in) or more; stipules ovate, to 1 cm ($^2/_5$ in) long. Flowers pink, often with paler centre; peduncle to 15 cm (6 in) with up to 15 flowers; upper petals with a few darker lines, obovate, *c.* 20 x 12 mm; lower petals slightly larger; hypanthium to 3 cm (1 $^1/_5$ in) or more; pedicel *c.* 5 mm; fertile stamens 7.

RECOGNITION: robust plant with unzoned, rounded leaves and pink flowers.

Although not very widely cultivated, this species is easy to grow and propagate and if it becomes too straggly, responds to cutting back to keep it more bushy. It will grow quite satisfactorily in slightly more shaded and damper conditions than many other pelargoniums which is explained by its natural habitat in the mountainous areas with higher rainfall in the Transvaal. Clifford suggested that it might be identical with *P. heterogamum* of L'Héritier but the older plant is different and probably an early zonal hybrid.

P. alchemilloides (Linnaeus) L'Héritier, Aiton, *Hortus Kewensis*, **2**: 419 (1789)

MEANING: resembles *Alcemilla* sp.

SECTION: **Ciconium**

ILLUSTRATION: *Pelargoniums of southern Africa*, **1**: 3.

Herbaceous, decumbent, hairy, perennial plant. Leaves, basically rounded, 5-lobed to varying depths, irregularly toothed, with cordate base, sometimes with a dark reddish-brown horseshoe-shaped zone, to 10 x 12 cm (4 x 4 $^4/_5$ in), often less; petiole to 10 cm (4 in); stipules ovate. Flowers cream, white or pale pink; peduncle to 20 cm (8 in), *c.* 5 flowered; upper petals sometimes with darker markings, spathulate, 10–20 x 2–10 mm; lower petals smaller; hypanthium to about 8 mm; pedicel *c.* 5 mm, strongly reflexed after flowering; fertile stamens 7, often with rather short filaments.

RECOGNITION: low-growing hairy plant with small white, cream or pink flowers, leaf often zoned.

Found over a very wide area of southern and eastern Africa, this species is exceedingly variable in flower and foliage. Several types are known in cultivation and yet others in the wild. A rather small form with zoned leaves, **P. ranuculophyllum** (Eckl. & Zeyh.) Baker,[5] is recognized as a distinct species and several other taxa may eventually be defined. The first record appears to be by Hermann in 1687 and it was cultivated by Jacob Bobart of the Oxford Botanic Garden as early as 1693.

P. ranunculophyllum

P. caylae Humbert, *Bulletin de la société botanique de France*, **82**: 595 (1935)

MEANING: after M. Cayla, Governor General of Madagascar

SECTION: **Ciconium**

ILLUSTRATION: *Pelargoniums of southern Africa*, **3**: 24.

Erect rarely branched shrub to over 1 m (40 in), with thick stems. Leaves with soft, short hair, heart-shaped, shallow-lobed, margins crenate, pale hairy below, to 8 cm (3 $^{1}/_{5}$ in) long; petiole *c.* 5 cm (2 in); stipules narrow ovate. Flowers slightly scented, pink-purple; peduncle many flowered, *c.* 15 cm (6 in); upper petals narrow obovate with undulate margins, *c.* 25 x 8 mm; lower petals slightly smaller; hypanthium 15 mm or more; pedicel *c.* 10 mm; fertile stamens 5.

RECOGNITION: tall, rarely branched shrub; flowers purple, petals wavy-edged; leaves resembling *P. cordifolium*.

If this species is planted in the ground rather than a pot, and allowed to grow upwards unimpeded, it will eventually flower. The velvety foliage is attractive but tends to be produced towards the top of a tall, bare, unbranched stem making it a curiosity and a challenge. The pelargonium collector will eventually be rewarded when it finally flowers. It is one of two species known from Madagascar being first discovered in 1934 growing in the mountainous regions of the south-east. (*Plate 4*)

P. elongatum (Cavanilles) Salisbury, *Prodromus Stirpium*, 312 (1796)

MEANING: elongated, presumably referring to the elongated flowering branches

SECTION: **Ciconium**

ILLUSTRATION: *Pelargoniums of southern Africa*, **1**: 44.

Straggling herbaceous plant to about 20 cm (8 in) or more, with long and short glandular hairs. Leaves reniform, 5–7 shallow-lobed, margins crenate, usually with dark-brown to purplish horseshoe-shaped mark at centre of lamina, 2–4 x 2–5 cm ($^{4}/_{5}$–1 $^{3}/_{5}$ x $^{4}/_{5}$–2 in); petiole to 7 cm (2 $^{4}/_{5}$ in) or more; stipules lanceolate with ciliate margin. Flowers small, normally cream-coloured; peduncle to 15 cm (6 in), 1–5 flowered; upper petals veined with red, narrow, *c.* 10 x 4 mm; lower petals slightly smaller; hypanthium 10–20 mm; pedicel 1–2 mm; fertile stamens 7.

RECOGNITION: zoned leaves with long hairy petioles and narrow stipules; flowers very small, pale yellow or cream.

This plant is short-lived but readily seeds itself everywhere. It is a rather insignificant plant which grows wild in south-west Cape Province and has been known at least since the Spanish botanist Cavanilles illustrated it in 1785. It is often found in cultivation labelled incorrectly as *P. tabulare* but Cavanilles recognized the difference between the two species and illustrates and describes both quite clearly. However, it was illustrated by L'Héritier as *P. tabulare* and this mistake was continued in some, but not all, later publications. *P. tabulare* is a separate pink-flowered species related to *P. patulum* in the section *Glaucophyllum*.

P. frutetorum R.A. Dyer *Kew Bulletin*, 446 (1932)

MEANING: shrubby
SECTION: **Ciconium**
ILLUSTRATION: Hooker, *Icones plantarum*, **32**: 3200 (1933).

Spreading, branching plant with thick reddish-brown stems. Leaves green with very distinct dark brown-purple ring-shaped zone towards centre of lamina, shallow 5-lobed, crenate, c. 6 cm (2 $^2/_5$ in) across; petiole c. 5 cm (2 in); stipules broad ovate, c. 10 mm. Flowers pale-salmon-pink; peduncle 10–15 cm (4–6 in), c. 10 flowered; petals obovate, 20 x 10 mm; hypanthium 4 cm (1 $^3/_5$ in); pedicel 2–4 mm; fertile stamens 7.

RECOGNITION: scrambling plant; leaf zoned; flowers pale-salmon-pink.

This species is superficially similar to *P. multibracteatum* but is less robust and has a much more clearly defined zone on the leaf. It grows in eastern Cape Province and was probably first collected by Burchell 180 years ago. It has been used in hybridization to create zonal type cultivars with attractive foliage and is tolerant of shade, a rare characteristic in the genus. 'The Boar' with a very dark brown central blotch instead of a zone to the leaf, was grown from plants found in Tresco on the Isles of Scilly. The plant received an Award of Merit at a show of the Royal Horticultural Society in 1956 as 'Salmonia' which has caused much confusion with *P. salmoneum*. At the time it was said to come true from seed although the dark blotch may take several months to appear. A similar plant with white flowers is 'White Boar', a seedling of 'The Boar'. *(Plate 5)*

P. 'The Boar'

P. inquinans (Linnaeus) L'Héritier, Aiton, *Hortus Kewensis*, **2**: 424 (1789)

MEANING: red, referring to the stain appearing when the plant is touched
SECTION: **Ciconium**
ILLUSTRATION: *Flowering plants of Africa*, **25**: 981; *Pelargoniums of southern Africa*, **1**: 23.

Erect branching, subshrub to 2 m (6 $^1/_2$ ft) velvety-hairy and with red glandular hairs. Leaves unmarked, almost circular, shallow-lobed, crenate, base cordate, to 8 cm (3 $^1/_5$ in) across; petiole to over 6 cm (2 $^2/_5$ in); stipules large, broad ovate to 15 x 20 mm. Flowers scarlet, occasionally pink or white, almost regular, buds reflexed; peduncle 10–20 flowered, to 8 cm (3 $^1/_5$ in); petals, broad ovate, 20 x 12 mm, upper 2 very slightly smaller than lower; hypanthium to 25 mm; pedicel 1–2 mm; stamens and style barely exserted, filaments joined for most of their length; fertile stamens 7.

RECOGNITION: erect shrub with red glandular hairs; flowers scarlet.

This species and *P. zonale* are generally considered to be the main parents of the modern zonal pelargoniums, often referred to as *P.* x *hortorum* Bailey. Both were very early introductions into the great gardens of Europe in the early eighteenth century. The first record appears to be of plants growing in the gardens of Fulham Palace, home of the Bishop of London, Dr Henry Compton in 1714, and the first illustrations in

Fig. 9 *P. inquinans* and *P. salmoneum* from Dillenius, *Hortus Elthamensis* (1732)

1732 taken from plants growing in James Sherard's garden not far away in Eltham. It is uncertain how this plant from the less explored eastern Cape Province was introduced but a few early expeditions were made in this direction and ships may also have stopped for fresh water on the coast near where the species grows. There are also accounts of an early exploration towards the east from Cape Town in 1689 under the leadership of Ensign Izaak Schryver, but whether or not plants were collected at this time is not recorded.

Many plants found under this name are early hybrids but the true species is easily recognized by the red glandular hairs found on the stems and leaves which, under most conditions, will leave a red-brown stain on the hands after touching the plant. This characteristic which appears to be lost during the first cross with another species, is very rarely mentioned but Linnaeus clearly describes it in 1753; 'folio digitis tacta inquinant colore ferrugineo'. The plant is another of those that has been used for its medicinal properties.
(Plate 6 and line drawing 9)

P. multibracteatum Hochstetter, *Flora*, **24:** 1 (1841)

MEANING: many bracts
SECTION: **Ciconium**
ILLUSTRATION: *Flowering plants of South Africa*, **80:** 794; *Pelargoniums of southern Africa*, **3:** 90.

Spreading, branching plant with thick stems to about 30 cm (12 in) tall. Leaves variable, faintly aromatic, pale-green, often with purplish zone half way to margin, softly-hairy, orbicular, lobed with crenate margin and cordate base, 5–7 cm (2–2 $^4/_5$ in) across; petiole 5 cm (2 in) or more; stipules triangular to 15 mm. Flowers white, tinged pink at base, almost regular; peduncle 5–10 flowered, 20 cm (8 in) or more, held well above the foliage; petals spathulate, $c.$ 20 × 10 mm; hypanthium 3–4 cm (1 $^1/_5$-1 $^3/_5$ in); pedicel 2–3 mm; fertile stamens 7.

RECOGNITION: flowers with broad white petals; leaves often zoned.

The variation in leaf form is not surprising when the wide natural distribution of this species through East Africa from Tanzania to Ethiopia is considered. It was first described in 1841 and although not widely known, is one of the few species with white flowers and soft green foliage. It is also adaptable to cultivation. **P. usambarense** Engl. and **P. hararense** Engl. with pink flowers are very closely related and plants from Yemen are similar in morphology.

P. mutans Vorster, *Flowering plants of Africa*, **52:** 2060 (1992)

MEANING: changeable referring to 4 or 5 petals
SECTION: **Ciconium**
ILLUSTRATION: *Flowering plants of Africa*, **52:** 2060.

Branching, sprawling plant with rather thick stems. Leaves aromatic, bright green normally with a purple zone about half way to margin, orbicular, 5 lobed to zone, margin dentate, base cordate, to 10 cm (4 in) across; petiole 3–4 cm (1 $^1/_5$–1 $^3/_5$ in); stipules broad ovate, 5 mm.

Flowers white to cream, very irregular; peduncle to 25 cm (10 in), c. 8 flowered; upper petals faintly lined, linear, reflexed, 20 × 5 mm; lower petals 2 or 3, straight, 15 × 3 mm; hypanthium 1–2 cm ($2/5$–$4/5$ in); pedicel to 1 cm ($2/5$ in); fertile stamens 7, almost equal, often curved upwards.

RECOGNITION: aromatic, zoned foliage; cream, very irregular flower with linear petals.

This is the plant which was been sold and grown incorrectly for many years as *P. grandiflorum* and was once included within the variable species, *P. alchemilloides*. It is not certain how it came to be in cultivation in Europe but was grown in European gardens for many years before it was collected in the wild in northern Natal in 1987 and eventually accepted as a distinct species. It is easy to propagate from seed or cuttings. The foliage is attractive with a rather pungent, but not sweet scent; the flowers cannot be described as striking.

P. peltatum (Linnaeus) L'Héritier, Aiton, *Hortus Kewensis*, **2**: 427 (1789)

MEANING: peltate referring to leaves

SECTION: **Ciconium**

ILLUSTRATION: *Botanical magazine*, **1**: 20; *Pelargoniums of southern Africa*, **I**: 33.

A trailing or climbing, variable perennial, the stems tending to be rather angular. Leaves somewhat fleshy, slightly aromatic, usually peltate, more or less rounded in outline, with 5 triangular lobes, 3–7 cm (1 $1/5$–2 $4/5$ in) across, glabrous and glossy or with a short soft velvety pubescence, often with a darker circular zone; petiole to 5 cm (2 in); stipules broad ovate. Flowers 2–9, strongly zygomorphic, white, pink or pale-purple, to 4 cm (1 $3/5$ in) across; peduncle c. 5 cm (2 in); upper petals 25 × 10 mm, spathulate, veined darker; lower petals shorter, unmarked; hypanthium 30–40 mm; pedicel c. 4 mm; fertile stamens 7, two very short.

RECOGNITION: trailing plant with ivy-shaped, often peltate, leaves.

Typically *P. peltatum* grows wild over a large area of the winter rainfall zone in Cape Province but the distribution extends much further east into the regions of summer rainfall of Natal and eastern Transvaal where the non-peltate leaved types are more frequent. It is usually found scrambling through other shrubs, able to reach up to 2 m (7 ft) or more.

It was apparently introduced into Europe in 1700 in a parcel of plants sent to Holland by the governor of the Cape Province. The rapid exchange of new plants at that time between the countries of Europe is indicated by the fact that it was being grown in England a year later by the Duchess of Beaufort who owned botanical gardens at Chelsea and Badminton in Gloucestershire. She also had the first stove house in the country which was used as a model for Queen Mary's.

The leaves have been used as an antiseptic and a deep-blue dye may be obtained from the petals which has been used in painting. In the wild, *P. peltatum* is quite variable in the size, texture and colouring of the foliage, some leaves being zoned while others are plain, some velvety in texture while others are smooth and shiny and some quite deeply divided while others are barely lobed. As the latin name suggests, the leaves are peltate, with the petiole attached to the centre of the leaf, but wild plants may also be found with leaves in which the petiole is attached to the edge of the leaf blade in a more usual manner. The flowers vary in size and colour from almost white to pale purple to pink. This variation has, in the past, led to the erroneous establishment of two or more distinct species or varieties but investigations[6] have shown that as there is a gradation between all forms in the wild, for botanical purposes, they should be included within the one species, *P. peltatum*. Very pubescent forms may still be found in gardens under the name *P. clypeatum*. Horticulturists like more exact parameters and so justifiably create many new names to distinguish the different cultivars. As long as there is understanding between the two disciplines and the nomenclatural rules of the botanical and horticultural worlds are respected, there need be no confusion or conflict.

The species is as easy as the ivy-leaved cultivars to both grow and propagate and the ivy-leaved pelargoniums of gardens today originated from the many variants of this species by both hybridization and sporting. *(Plate 7)*

P. 'Lateripes'

MEANING: stalk on side of leaf
SECTION: **Ciconium**
ILLUSTRATION: L'Héritier, 24; Andrews, *Geraniums*, **1**: 1805.

Very similar to *P. peltatum* except that the petiole is joined to the edge of the leaf blade, the stem is rounded and the habit more erect. There may be more flowers in each head and the flowers are larger, paler-pink with reddish-purple feathering on the upper petals. The stamens are usually deformed or aborted and do not produce viable pollen. As early as the end of the eighteenth century there are records of variegated forms of this plant and it still mutates readily to produce plants with differing flower colours or with extra petals.

Confusion arises because plants with similar foliage are found wild and are rightly included within the variable species, *P. peltatum*. However the original plant cultivated under this name was most likely a hybrid. Both L'Héritier in 1789 and Andrews in 1805 illustrated a purplish-pink-flowered ivy-leaved pelargonium. L'Héritier called his plant *P. lateripes* and implied that he considered it might be of hybrid origin which the rather small anthers in his illustration would infer. That of Andrews was raised, or at least cultivated, by Messrs Grimwood and Barret in about 1787. However, Andrews confuses the issue by deliberately changing the name to *P. hederinum* using an adaptation of the latin name for ivy, *Hedera*, as a specific name to match the vernacular english name, ivy-leaved. He then distinguished *P. peltatum* as the 'old ivy-leaved pelargonium'. As the plant has been cultivated for nearly two hundred years as *P. lateripes*, it is permissible in this case, under the rules on the nomenclature for cultivated plants to use the latin epithet as the cultivar name when changing the status of the plant from species to cultivar.

P. quinquelobatum Hochstetter, Richard, *Tentamen Florae Abyssinicae*, **1**: 118 (1847)

MEANING: 5-lobed leaf
SECTION: **Ciconium**
ILLUSTRATION: *Flowering plants of South Africa*, **20**: 795.

Herbaceous, hairy, perennial plant. Leaves dull-green with bluish tinge, broad-triangular in outline with deep cordate base, 5-lobed to more than half way to midrib, lobes triangular with large irregular teeth, to about 10 x 8 cm (4 x 3 $^1/_5$ in); petiole 8–12 cm (3 $^1/_5$–4 $^4/_5$ in); stipules lanceolate to 1 cm ($^2/_5$ in). Flowers pale yellowish-green to greyish-green-blue; peduncle to 30 cm (12 in) bearing about 5 flowers; upper petals faintly lined with pink, obovate, *c.* 15 x 5 mm; lower petals slightly broader, unmarked; hypanthium to 30 mm; pedicel to 5 mm; fertile stamens 7, 3 with very short filaments.

RECOGNITION: curious grey-blue-yellow flowers.

The species is found from Tanzania and Kenya to Ethiopia and plants collected from different localities have been accorded different names. The more southerly growing *P. fischeri* Engl. from Tanzania is almost identical in cultivation, but the flower is slightly larger and has a more yellow colour; it is best considered as a variation of *P. quinquelobatum*. There must be several pigments in the petals to create the unusual shades of colour which have proved almost impossible to reproduce accurately on film. At about the turn of the century, a form with greenish-yellow flowers was exhibited before the Royal Horticultural Society as *P. inaequilobum*. This was being used for hybridizing at the time but there are no obvious results remaining in cultivation today. Perhaps the colours are equally difficult to introduce into other pelar-goniums. One in cultivation is the pink-flowered 'Pink Quink', a cross with a dwarf zonal, 'Clarissa' made by Faye Brawner in USA. *(Plate 8)*

P. salmoneum R.A. Dyer, *Kew bulletin*, 447 (1932)
MEANING: salmon, referring to flower colour
SECTION: **Ciconium**
ILLUSTRATION: *Botanical magazine*, **157**: 9357; *Flowering plants of Africa*, 971.

Weakly erect, branching slightly glandular hairy, subshrub to about 1.5 m (5 ft) or more with semi-succulent stems. Leaves thick, somewhat folded upwards, green to somewhat glaucous, unmarked, reniform with truncate or shallow cordate base, margin crenate, to 4.5 × 5 cm (1 ^{4}/$_5$ × 2 in); petiole to 4 cm (1 ^{3}/$_5$ in); stipules broad ovate to 15 mm. Flowers salmon-pink, buds strongly reflexed; peduncle to 15 cm (6 in), to 15 flowered petals almost equal, narrow obovate, *c.* 2.5 × 1.5 cm (1 × ^{3}/$_5$ in), upper 2 veined darker; hypanthium to 25 mm; pedicel to 5 mm; fertile stamens 5.

RECOGNITION: resembles *P. acetosum*, but leaves slightly greener, not cuneate and flowers with broader petals.

P. salmoneum was first found in a garden in Port Elizabeth and described in 1932. For many years it was considered to be a species but later, because it was only known in cultivation, it was listed as a cultivar. However, it behaves as a species, is easily reproduced by seed and presumed to be native to eastern Cape Province where several closely related plants may be found in the wild. Herbarium specimens at Kew taken from material cultivated by Thomas Moore in 1867 at Chelsea as *P. roseum*, look very similar but even more interesting is the illustration in Dillenius of 1732 of a plant that appears far closer to this species than *P. inquinans* under which it is listed as a variety. The description of Dillenius is of a plant with glaucous leaves and pale-red flowers similar in habit to *P. acetosum*. Surely this description equates with the *P. salmoneum* we know today? Early authors such as Sweet (**1**: 63) suggested that the plant illustrated by Dillenius was a hybrid of *P. zonale* and *P. inquinans* and named it *P. hybridum*, but this latter plant had a completely different habit and scarlet flowers. This species should not be confused with the name 'Salmonia' originally used incorrectly as a synonym for 'The Boar', a cultivar related to *P. frutetorum*. *(Fig. 9)*

P. 'Saxifragoides'
MEANING: resembling a saxifrage
SECTION: **Ciconium**
ILLUSTRATION: Clifford, fig. 28.

A low-growing, spreading, hairless plant. Leaves, tiny, sometimes peltate, usually similar in shape to *P. peltatum* or sometimes somewhat deformed, to about 2 × 2.5 cm (^{4}/$_5$ × 1 in) but often smaller, thick and fleshy; petiole to over 2.5 cm (1 in); stipules minute. Flowers 2–5, pinkish-white, star-shaped, under 1 cm (^{2}/$_5$ in) across; upper petals, oblong, reflexed, 5 × 3 mm, with darker markings; lower petals narrower, spreading, unmarked; hypanthium *c.* 2 mm; pedicel to 4 mm; stamens deformed.

RECOGNITION: plant with small ivy-shaped leaves and small star-shaped pink flowers.

This plant was originally described as a species at Kew[7] but is more likely to be a hybrid or mutant of *P. peltatum*. The stamens are abortive and it rarely, if ever, produces any viable seed. It was found in the Royal Horticultural Society's garden at Chiswick before 1890. Its original source is unknown but it may have been imported unknowingly with other plants. Kew obtained the plant soon after. Another very similar plant said to have been discovered by Bachmann in Pondoland in 1888, was described by Knuth as a new species under the name *P. bachmannii*, but unfortunately, there is no description of the flowers and the original herbarium specimens have been destroyed. However it seems highly likely that it is the same as *P.* 'Saxifragoides' and might have been the source of the plant found at Chiswick. Probably no one will ever know its true history for certain.

P. stenopetalum Ehrhart, *Beiträge zur Naturkunde*, **7**: 161 (1792)

MEANING: narrow petals
SECTION: ***Ciconium***

This is taxonomically another problem plant. In appearance the flowers are quite distinct with deep coral to scarlet-red petals which are narrower than the sepals. The upper two are held erect and in close contact for half their length and then diverge at a wide angle ending in pointed, not rounded tips, while the lower three curve downwards. The faintly zoned leaves have a shallow cordate or truncate base and the plant has a scrambling untidy habit similar to but less vigorous than *P. salmoneum*.

In 1792, Ehrhart first described a plant with scarlet flowers, unmarked petals narrower than the sepals and a zoned leaf. Later books of the early and mid nineteenth centuries list this plant. Unfortunately, the illustration in Andrews under this name shows a plant with shorter spreading petals which Sweet later refers to as *P. leptopetalum*.[8] It would seem likely therefore that there were two plants in circulation and that *P. stenopetalum* of Ehrhart is what has been grown, although never commonly, for 200 years in Europe under this name. This plant may have been involved in early attempts at hybridization but as the older gardeners were interested in creating more substantial flowers with rounded petals and a fuller inflorescence, it was probably ignored unless the bright colour was needed.

In 1927[9] a plant was described and named *P. burtoniae* from a specimen found by a Mrs Burton in a garden in eastern Cape Province in South Africa. It has more recently been collected in Natal but is identical with the plant grown in Europe as *P. stenopetalum*. Whichever is the correct name, the plant is instantly recognizable and the flower is quite unique. *(Plate 9)*

P. tongaense Vorster, *South African journal of botany*, **2**: 76 (1983)

MEANING: area of Natal where first discovered
SECTION: ***Ciconium***
ILLUSTRATION: *Pelargoniums of southern Africa*, **3**: 141.

Semi-erect hairy, herbaceous plant. Leaves with 3–5, coarsely toothed, triangular lobes, thick to somewhat fleshy, 4–7 cm (1 $^3/_5$–2 $^4/_5$ in) across; petiole to 10 cm (4 in); stipules ovate. Flowers bright red, to 2 cm ($^4/_5$ in) across; peduncle to 20 cm (8 in) bearing up to 10 flowers; upper petals ovate, *c.* 17 × 7 mm; lower petals slightly broader; hypanthium to 4 cm (1 $^3/_5$ in); pedicel very short; fertile stamens 7, filaments joined for most of their length.

RECOGNITION: somewhat similar to *P. peltatum* in foliage but flowers bright red.

Although only named in 1983, it had been known in South Africa for nearly thirty years and was introduced to the gardens of the RHS at Wisley in 1972, as 'seed collected in Pongoland'. In cultivation, it is a very attractive and tolerant plant, surviving in shade or sun and seeds freely.

In the wild it grows in humid conditions in shaded areas in Natal.

P. transvaalense Knuth, *Das Pflanzenreich* 4, 129: 434 (1912)

MEANING: from Transvaal
SECTION: **Ciconium**
ILLUSTRATION: *Pelargoniums of southern Africa*, **2:** 149.

Plant with short erect red-tinged stems. Leaves hairy, often zoned, with 5 triangular, dentate lobes, *c.* 8 x 8 cm (3 $^1/_5$ x 3 $^1/_5$ in); petiole tinged red, to 10 cm (4 in); stipules narrow ovate. Flowers purple pink, large; peduncle 5–10 flowered, to 10 cm (4 in); upper petals veined deeper red, spathulate, apex notched, 18 x 6 mm; lower petals narrower; hypanthium 2 cm ($^4/_5$ in); pedicel *c.* 12 mm; fertile stamens 7.

RECOGNITION: similar to but more robust than *P. alchemilloides* in habit but with flowers resembling *P. zonale*.

This is another species that grows in wooded, damp situations in eastern Transvaal, and which adapts well to cultivation. It is of relative recent introduction being first described early this century. Some of the early accounts appear to be of *P. multibracteatum*. It was used as a cure for intestinal disorders and fevers.

P. zonale (Linnaeus) L'Héritier, Aiton, *Hortus Kewensis*, **2:** 424 (1789)

MEANING: zoned, referring to leaf
SECTION: **Ciconium**
ILLUSTRATION: *Pelargoniums of southern Africa*, **1:** 50.

Erect usually hairy subshrub to about 1 m (40 in) but often found scrambling through bushes or cascading down cliffs. Leaves orbicular, crenate, often with a darker brownish-purple horseshoe-shaped mark, 5–8 cm (2–3 $^1/_5$ in) across; petiole to 5 cm (2 in); stipules large, brown membranous. Flowers usually pale pink, sometimes white or red, buds reflexed; peduncle to 20 cm (8 in), to 50 flowered, but usually less; petals more or less equal in size, oblanceolate, marked with darker veins, upper 2 erect, lower 3 spreading, *c.* 20 x 7 mm; hypanthium to over 25 mm; pedicel 1–2 mm; fertile stamens 7, 2 very short.

RECOGNITION: robust scrambling plant; leaves often zoned; inflorescence many flowered, petals narrow.

Commelin, records how this species was grown in 1700 in the medicinal garden of Amsterdam, from seed sent to Holland by the governor of the Cape of Good Hope, Wilhelm Adrian van der Stel. He describes the zoning on the leaves and the reddish-coloured flowers. Four years later it was recorded in the garden of the Duchess of Beaufort, who had gardens at Chelsea and Badminton in England, and it has been in cultivation ever since. This is yet another example of the close relationship between the two countries at this time and the speed of exchange of botanical knowledge. In the wild, it is widespread from eastern to western Cape Province and the robust

Fig. 10 *P. zonale* from Commelin, *Icones plantarum horti medici Amstelaedamensis*

scrambling plants would have been an obvious choice for early plant collectors as well as gardeners. Although the majority of plants seen wild today have pinkish-coloured flowers, sometimes with only faintly zoned leaves, several of the references specifically mention red flowers with zoned leaves so perhaps these were the ones selected. It has lent its name to the enormous group of modern cultivars, the zonals, and was certainly involved in their parentage. The fuller rounder flower which was the aim during the early days of hybridization would have come from other species such as *P. inquinans*. (Plate 10 and fig. 10)

P.* × *kewense R.A. Dyer, almost certainly a hybrid of *P. zonale* and *P. inquinans*, was found at Kew and named in 1932. It has an inflorescence with up to 27 bright crimson-red flowers with narrow to obovate petals. The shallow-lobed leaves are usually unmarked, but sometimes have a faint zone when young. The base is shallow, cordate and the margins crenate[10] One or two of the early Sweet illustrations of zonal type hybrids are not dissimilar.

P. 'Scandens'

This cultivar is probably of similar parentage to *P.* × *kewense* but the larger leaves have a narrow, very distinct and consistent zone towards the edge of the lamina and the leaf base is cordate. The flowers are bright red with narrow petals. The exact nomenclature of this plant although commonly grown, is subject to question. The name was first used by Ehrhart and by later authors to describe a plant with rather narrow petals of pale pink or white flushed with pink, a zoned leaf and a somewhat climbing habit. In the wild, forms of *P. zonale* would match this description and it is almost certain that *P. scandens* Ehrh. was one of these.

Cortusina (de Candolle) Harvey

Until quite recently, this was a much larger section including plants which have now been transferred[11] to the new section, *Reniformia*. Although there are similarities between the two, the many distinct morphological characteristics as well as the very widely separated geographical locations make this an obvious division. In 1824, de Candolle created a subseries for species such as *P. echinatum*. Ecklon and Zeyher raised it to the status of genus in 1835 but Harvey in *Flora Capensis* in 1860 used the name as a section. Retained within the limits of the present section *Cortusina* are the plants from the desert and semi-desert areas of the north-western part of Cape Province and Namibia where many may be found growing in the shelter of rock crevices, thus gaining some protection from the harsh conditions.

The rainfall is exceedingly low, often under 10 cm (4 in) annually. What does fall is in winter but the plants will obtain additional moisture from condensation forming on the leaves overnight and the fogs which blow in off the sea. The days, especially in summer, can be very hot and at night the temperature will drop considerably so the plants must be able to withstand great extremes of temperature. The stems of all except one, are succulent for the storage of water and often covered with the remains of persistent hard stipules or petiole bases, giving extra protection. The long petioled, simple, usually almost rounded leaves, are shed in periods of drought, also to conserve water by reducing transpiration. Several species have the added advantage of tuberous or thickened roots which again aid survival in extreme conditions.

However, as well as including some plants with rather curious shapes, this section contains species with some of the most attractive flowers of the genus making them worth growing even though for much of the year the plants are dormant and rather uninspiring. In the wild they flower in spring after the rain, over a fairly short time span, presumably an adaptation to the short favourable season. In cultivation they also bloom early in the year. The inflorescence is rarely branched, but each peduncle bears several flowers. Each flower has an almost regular appearance with 6 or 7 fertile stamens and, except for *P. desertorum*, the hypanthium is always conspicuously longer than the pedicel. The basic chromosome number is $x = 11$ and most species have 22 fairly small chromosomes. Hybrids are almost unknown within this section although the cultivar 'Miss Stapleton' may be one example. If the nomenclature used in the days of Sweet is unravelled and a systematic programme of hybridization is undertaken, it may be that some of his many illustrations of new plants will prove to be the result of crosses of species within this section.

The flowers and simple leaf show some resemblance to *P. cotyledonis* but these similarities are probably coincidental and do not necessarily mean a close relationship. The section shows some relationship with the sections *Pelargonium* and *Glaucophyllum* but except in some morphological features, little with recently spilt-off section *Reniformia*. Some species of *Ligularia* resemble those of *Cortusina*, and *P. desertorum* and *P. xerophyton* have been moved from *Ligularia* fitting neatly within the boundaries of the newly redefined section, *Cortusina*.

The plants are usually easy to grow as long as their natural habitat is considered and they are given plenty of light, a very free-draining soil and are never overwatered. The leaves may not drop, and the plants continue to grow when cultivated under glass or in less extreme conditions than those experienced in their native habitat. Under cultivation however it is advisable to reduce watering considerably during the period of less active growth. Allowing the plants to rest for part of the year appears important for maintaining their vigour, and in the more temperate climates of Europe and North America this is usually in the winter when the light is less intense. Cuttings strike readily during the growing period and seed is often formed which, although it may take several months, can be germinated successfully.

Some of the species are among the earlier ones to be discovered during explorations travelling north from the Cape of Good Hope, but although *P. echinatum* has been grown continually since its first introduction in 1794, many others were lost to cultivation and only reintroduced quite recently.

1	Plant upright, shrubby	2
	Plant without an aerial stem	**sibthorpiifolium**
2	Stems spiny	**echinatum**
	Stems not spiny	3
3	Flowers bright-pink	**magenteum**
	Flowers mauve or white, not bright-pink	4
4	Leaves green, usually under 20 mm long	5
	leaves grey-green, usually over 30 mm long	6

5	Leaves strongly aromatic; hypanthium not exceeding pedicel	**desertorum**
	Leaves not or faintly aromatic; hypanthium exceeding pedicel	**xerophyton**
6	Leaf base tapering into petiole	**crassicaule**
	Leaf base heart-shaped	**cortusifolium**

P. cortusifolium L'Héritier, Aiton, *Hortus Kewensis*, **2**: 428 (1789)

MEANING: leaves resemble *Cortusa* sp.

SECTION: **Cortusina**

ILLUSTRATION: Sweet, **1**: 14; *Pelargoniums of southern Africa*, **2**: 36.

Branching plant to 30 cm (12 in) with thick succulent stems covered with the remains of membranous stipules. Leaves with a silvery sheen, hairy, almost rounded, shallow-lobed, base cordate, to 4 cm (1 $^3/_5$ in) across; petiole 2–8 cm ($^4/_5$-3 $^1/_5$ in); stipules narrow triangular. Flowers white to pale pink or lilac in inflorescence of about 10 flowers; peduncle to 4 cm (1 $^3/_5$ in); upper petals spathulate, often with dark-red markings, to 14 x 8 mm, reflexed; lower petals very similar to upper but often darker in colour; hypanthium 20–30 mm; pedicel 1–2 mm; fertile stamens 6.

RECOGNITION: similar to *P. crassicaule* but distinguished by cordate leaf base and persistent stipules resulting in a rougher stem texture.

This species is found along the coastal areas of southern Namibia in desert conditions. The cool sea currents ensure that the summers are cooler than might be expected for the region and sea mists supplement the very low annual rainfall. It was first collected by A.P. Hove in 1786 during an expedition to the south-west coast of Africa. Like many other species, it was well-known to the *Pelargonium* collectors of the early nineteenth century and like *P. crassicaule*, illustrated by all the major botanical artists of the time. Unfortunately it was overlooked in the enthusiasm of later years for hybridizing and the creation of new and large-flowered cultivars. Like *P. crassicaule*, it appears slightly more tender than some of the other species.

P. crassicaule L'Héritier, Aiton, *Hortus Kewensis*, **2**: 428 (1789)

MEANING: thick stem

SECTION: **Cortusina**

ILLUSTRATION: *Botanical magazine*, **14**: 477; Sweet, **2**: 192; *Flowering plants of South Africa*, **2**: 52; *Pelargoniums of southern Africa*, **2**: 41.

A low-growing plant with thick brown knobbly succulent stems. Leaves grey-green with a soft velvety texture, very broad ovate, base cuneate, margins crenate, to 5 cm (2 in) across; petiole 3–5 cm (1 $^1/_5$–2 in); stipules triangular, semi-persistent. Flowers slightly scented, white, to pale pink or lilac in inflorescence of about 8 flowers; peduncle to 4 cm (1 $^3/_5$ in); upper petals spathulate, often blotched and lined with purple, reflexed, to 15 x 7 mm; lower petals similar but slightly smaller, often darker than upper; hypanthium 15–20 mm; pedicel 1–2 mm; fertile stamens 6 or 7.

RECOGNITION: similar to *P. cortusifolium* but distinguished by cuneate leaf base.

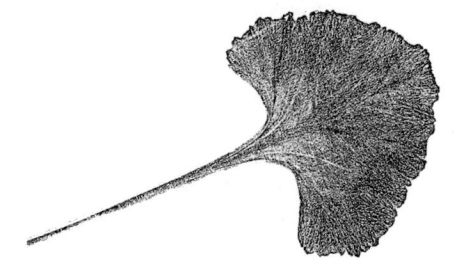

Like *P. cortusifolium*, this species was introduced by Hove in 1786 from southern Namibia where it grows in similar conditions along the coastal strip. Its natural distribution does however extend further south into the north-western part of the Cape Province. Plants growing further inland with dark purple stems and less hairy leaves may be a form of the species or perhaps equivalent to *P. mirabile* Dinter, but these are not apparently in cultivation in Europe. The species was well-known in the heyday of *Pelargonium* species in the early nineteenth century when forms were known with large dark purple spots on each petal or with very faint spots. Sweet noted that to retain the dark spotted one, which was considered more desirable but less easy to propagate, it had to be self-pollinated. Neither form is common in cultivation today.

P. desertorum Vorster, *South African journal of botany*, **52**: 184 (1986)

MEANING: desert, referring to the natural habitat
SECTION: **Cortusina**
ILLUSTRATION: *Pelargoniums of southern Africa*, **3**: 42.

Small erect branching plant to 30 cm (12 in), with thin succulent stems. Leaves aromatic, bright green, rounded, shallow-lobed, base extending into petiole, 1–2 cm ($^2/_5$–$^4/_5$ in) across; petiole to 2 cm ($^4/_5$ in), semi-persistent; stipules triangular, persistent. Flowers white; peduncle 2–3 cm ($^4/_5$–1 $^1/_5$ in), 2–4 flowered; sepals reddish-purple-tinged; upper petals held erect, narrow spathulate, with red-purple lines and flushed reddish at base, 12 × 4 mm; lower petals similar but wide spreading; hypanthium variable but never exceeding pedicel; pedicel 2–3 cm ($^4/_5$–1 $^1/_5$ in); fertile stamens 7.

RECOGNITION: a small subshrub distinguished from *P. xerophyton* by its aromatic foliage and smaller but more numerous flowers.

This species was not knowingly collected until 1975 but being similar to *P. xerophyton*, it may have been seen many years ago, remaining unrecognized until the detailed study of the botanists at Stellenbosch University. It is found in a small area near the lower end of the Orange River in north west Cape Province, where the temperatures are very high in summer but cool at night. It is an interesting plant for the gardener as it tends to flower over a long period and is not difficult to cultivate.

P. echinatum Curtis, *Botanical magazine*, **9**: 309 (1795)

MEANING: spiny, referring to stems
SECTION: **Cortusina**
ILLUSTRATION: Sweet **1**: 54; *Pelargoniums of southern Africa*, **1**: 13.

Erect subshrub with thick spiny stems and thickened roots. Leaves ovate, shallow 3–5 lobed, crenate, grey-green, paler and hairy below, to 6 cm (2 $^2/_5$ in) across; petiole 10 cm (4 in); stipules awl-shaped, persistent, becoming hard and spiny. Flowers usually white in cultivation but also pink to purple in wild, 1.5–2 cm ($^3/_5$–$^4/_5$ in) across; peduncle 3–8 flowered, 3–4 cm (1 $^1/_5$–1 $^3/_5$ in); upper petals each with a dark-red blotch and lines, ovate, *c.* 20 × 10 mm; lower petals smaller, usually unmarked; hypanthium 3–4 cm (1 $^1/_5$–1 $^3/_5$ in); pedicel 3–4 mm; fertile stamens 6–7.

RECOGNITION: thickened spiny stems.

In the wild, *P. echinatum* has variable-coloured flowers, and many of the plants described and illustrated as new species or hybrids in the early books on the genus may simply be variations of the species. In some early catalogues, as many as eight different forms were listed. The white-flowered form of the species is sometimes known by the common name, the sweetheart pelargonium on account of the heart-shaped red blotches on the petals, and incorrectly by the name *P. echinatum* 'Album' to distinguish it from 'Miss Stapleton'. It would appear that 'Miss Stapleton' which has bright purplish-pink flowers spotted on each petal and longer straighter spines, (Sweet, **3**: 212), is the only named hybrid remaining today. For many years it was thought to be a form of *P. echinatum* or a hybrid of uncertain parentage but there are suggestions that it is the result of a cross between *P. echinatum* and *P. cortusifolium*. It was raised from seed collected at Colvill's nursery and first flowered in 1823, and named for Miss C. Stapleton, 'a lady much attached to the Geraniaceae'. The nursery must have been quite pleased with this plant because Sweet adds:

> We understand we are very much envied in a certain quarter for raising so many beautiful hybrid plants and more so for publishing them: but we mind not their envy, as long as we are so ably supported by our numerous subscribers to whom we beg our most grateful acknowledgements, and who we shall always use our utmost endeavours to please.

He would have been astonished at the price paid for his works today!

A white-flowered specimen was first collected by F. Masson for Kew in 1789 and unlike many others has remained in cultivation ever since. This is probably because it is not quite so exact in its cultural requirements as some other members of the section. It grows on dry ground often in the shelter of rocks or larger plants in the north-western regions of Cape Province. *(Plate 11)*

P. magenteum J.J.A. van der Walt, *Journal of South African botany*, **46:** 284 (1980)

MEANING: bright-pink, referring to flowers
SECTION: **Cortusina**
ILLUSTRATION: *Pelargoniums of southern Africa*, **1:** 36; *Flowering plants of Africa*, **49:** 1947.

Shrub-like plant to 90 cm (35 in) with woody branches. Leaves grey-green, rounded to reniform, hairy, crenate, to 15 mm or more across in cultivation but less in wild plants; petiole 1–2 cm (^{2}/$_5$–^{4}/$_5$ in); stipules broad-triangular. Flowers bright-pink with lines and a large, very dark blotch on each petal; peduncle 3–9 flowered, to 5 cm (2 in); upper petals broad spathulate, *c.* 2.5 x 1.5 cm (1 x ^{3}/$_5$ in); lower petals oblong, 2 x 1 cm (^{4}/$_5$ x ^{2}/$_5$ in); hypanthium to 3 cm (1 ^{1}/$_5$ in); pedicel 1–3 mm; fertile stamens 6–7.

RECOGNITION: bright pink flowers with spots on each petal.

This species could initially be mistaken for 'Miss Stapleton' with the bright-pink flowers and spots on each petal and it may have been involved in the parentage of this cultivar. It is however much less succulent and the woody stems are not spiny. It has been known as *P. rhodanthum* Schltr. but should not be confused with the plant illustrated by Sweet (**3:** 291) under this name which is a quite unrelated early hybrid. It grows in the drier more northerly inland parts of south-western Cape Province further south than other species of the section. It would appear that the early illustrations of *P. reniforme* (*Botanical magazine* 493 and Sweet plate 48) are very close to this plant and probably the references to the collection by Masson in 1791 also refer to this species. This would make more sense as the natural distribution of pelargoniums collected in the same year matches the distribution of this species but not of *P. reniforme* as is known today. It has been used medicinally for dysentery and other intestinal complaints.

P. sibthorpiifolium Harvey, *Flora Capensis*, **1:** 301 (1860)

MEANING: leaf similar to those of *Sibthorpia*
SECTION: **Cortusina**
ILLUSTRATION: *Pelargoniums of southern Africa*, **2:** 131.

This plant has flowers resembling *P. cortusifolium* but the reniform leaves are borne at soil level on a very extensive underground stem. It has several tubers and dies down completely in unfavourable conditions so is almost impossible to locate except for short periods each year. During the explorations of Paterson and Gordon, northwards along the west coast in 1779, this plant was obviously noticed as it is clearly recognizable from an illustration in the account of their journeys.

About fifty years later it was found again but has rarely been cultivated in Europe. It grows in the desert conditions of coastal areas to the north and south of the Orange River where the low-growing habit and underground tubers enable it to survive. It is an unusual species and if obtainable is worth attempting to cultivate as a curiosity. Like many tuberous plants it is extremely sensitive to overwatering. The very free-draining mixture of equal quantities of horticultural sand and soilless compost, as well as feeding with a liquid fertilizer and situating in strong direct light might help maintain this rather difficult species.

P. xerophyton Schlechter ex Knuth, *Das Pflanzenreich 4*, **129:** 383 (1912)

MEANING: a plant of arid habitats
SECTION: **Cortusina**
ILLUSTRATION: *Pelargoniums of southern Africa*, **3:** 149.

A low-growing plant to about 30 cm (12 in) high in the wild but larger in cultivation with numerous thick and almost woody branches. Leaves dull to blue-green, broad ovate, toothed in upper part, base cuneate, to about 10 mm across; petiole to 10 mm; stipules triangular. Flowers white usually solitary on peduncle to 15 mm; sepals green; upper petals narrow spathulate, with red spots and lines, to 20 x 5 mm; lower petals slightly smaller, unmarked; hypanthium 10–25 mm; pedicel 1–2 mm; fertile stamens 7.

RECOGNITION: similar to *P. desertorum* but leaves barely aromatic, and flowers larger but fewer.

Native to the very hot, dry desert areas of southern Namibia and north-western Cape Province, this species survives in the shelter of rocks as compact, low-growing, plants although in the more favourable conditions under cultivation will become larger more open and upright. It was first recorded in 1897 but is rarely grown except by some collectors of succulents or enthusiasts of the genus. *(Plate 12)*

Glaucophyllum Harvey

The leaves of the species of this rather small section are usually, as the name suggests, glaucous, and often rather leathery or succulent, simple or divided. The petioles often appear to be jointed at the base. The white, cream or purplish-pink flowers in a few flowered inflorescence, are very irregular in shape with a hypanthium usually considerably longer than the pedicel. Five to seven fertile stamens are present. The plants are subshrubs, either upright or trailing and adapted to live in the rocky, mountainous regions of southern, and south-western parts of Cape Province in very free-draining soils. Here they experience conditions of high light intensity, hot usually dry summers and cold winters during which time most of the rainfall occurs.

As with any other *Pelargonium* species, if care is taken to avoid waterlogging, species of this section are not difficult to grow and may be propagated easily by seed or cuttings. They are similar in many morphological features to species of the section *Pelargonium*, but show characteristics adapted to the drier more arid conditions in which they grow. Natural hybrids found in the wild where the two sections grow in association, as well as the several known artificial hybrids between members of the two sections also indicate their close relationship. The chromosomes are similar and small with the same basic number of $x = 11$. All except some varieties of *P. patulum* are diploid.

The section was first delimited by Harvey based on the glaucous succulent foliage but a recent comprehensive revision[12] has removed some species and added others.

1. Leaves lanceolate, not lobed or divided — **lanceolatum**
   Leaves not lanceolate, always lobed or divided — 2
2. Leaves simple although sometimes deeply divided — 5
   Leaves compound — 3
3. Fertile stamens always 5; leaf subtending peduncle with petiole under 5 mm long — **ternatum**
   Fertile stamens 5, 6 or 7; leaf subtending peduncle with petiole over 5 mm long — 4
4. Leaves divided into obovate segments not rounded on under side — **fruticosum**
   Leaves divided into linear segments rounded on under side — **laevigatum**
5. Plant decumbent; leaves on inflorescence sessile — **patulum**

Plant erect; leaves on inflorescence
stalked 6
6 Petals over 15 mm long; flowers
white, pink or pale purple **grandiflorum**
Petals under 15 mm long;
flowers pink **tabulare**

P. fruticosum (Cavanilles) Willdenow, *Species plantarum*, **3:** 689 (1800)

MEANING: shrub-like
SECTION: **Glaucophyllum**
ILLUSTRATION: *Pelargoniums of southern Africa*, **2:** 63.

Superficially similar to *P. laevigatum* in general appearance but with a more untidy and open habit; leaves less succulent and less glaucous, divided into linear flattened segments, not rounded on the under side; stipules broad-triangular; fertile stamens 5.
Botanically, this species is closer to *P. ternatum* the leaves of which have much broader leaflets. It is found wild over quite a wide area in varying habitats in southern Cape Province. Although known by European botanists since the end of the eighteenth century, this species has never been a popular plant as the small flowers are rather intermittent.

P. grandiflorum (Andrews) Willdenow, *Species plantarum* **3:** 674 (1800)

MEANING: large flowers
SECTION: **Glaucophyllum**
ILLUSTRATION: *The botanist's repository*, **1:** t. 12; Sweet, **1:** 29; *Pelargoniums of southern Africa*, **1:** 19.

An erect, almost hairless, shrubby plant, woody at base to 50 cm (20 in) or more. Leaves glaucous, sometimes zoned, deeply 5–7 palmately lobed, lobes coarsely toothed, to 5 cm (2 in) across but usually less; petiole about length of lamina; stipules broad ovate to 15 mm or more. Flowers 2–3, pink to pale purple-pink in most plants seen or creamy-white, upper petals about 20 x 12 cm (8 x 4 $^4/_5$ in), or more, narrow obovate, marked with darker blotches and streaks, lower narrower, unmarked; to 4 cm (1 $^3/_5$ in) or more bearing smaller stalked leaves; hypanthium *c.* 10 mm; pedicel *c.* 10 mm; fertile stamens 7.

RECOGNITION: upright plant with glaucous, maple-shaped leaves and large pink (or white) flower.

In the wild it may be found in mountainous regions of south-western Cape Province.

Although it has been listed frequently in nursery catalogues, the plant described above is rarely seen. For several years, most plants listed as *P. grandiflorum*, at least in Europe, were the recently named *P. mutans*. There seem to be no plants nowadays resembling the first illustrations, descriptions and herbarium specimens of the late eighteenth and nineteenth centuries with very large white flowers marked with purple veins.

In Andrews *The botanist's repository (plate 12)*, is the first colour illustration of a large white-flowered plant with upper petals veined purplish-red, to 35 x 18 mm and a hypanthium to 3 cm (1 $^1/_5$ in), three to four times longer than the pedicel. The cordate glaucous leaf was not zoned and the stipules were not very obvious. It was said to have been introduced in 1794 by F. Masson. A specimen dated 1797 in the Kew herbarium matches this illustration as does that of Sweet (**1:** 29) and all the descriptions by numerous authors up to the time of Knuth in *Das Pflanzenreich* in 1912. This white-flowered plant of the early nineteenth century was used for breeding. It has been thought for many years that it played a significant part in the ancestry of the regal pelargoniums.

Sweet noted in 1820 that although the species itself was becoming scarce, there were a great number of hybrids of it. He illustrates several raised by the enthusiasts of the day whose aim would have been to produce larger, more showy flowers. For example in Sweet (**5:** 45) is an illustration of a hybrid with *P. cucullatum* raised in 1821 which has large, white, rounded cup-like petals resembling a forerunner of an early regal type.

Andrews depicts an even larger-flowered hybrid named 'Formanii' after the raiser at Oxford Botanic Garden, and several specimens in the Kew Herbarium match this illustration with truncate and more closely-toothed leaves. As well as this, he also describes a pink-flowered plant although it was thought at the time to be a cultivated form. De Candolle describes this latter plant as 'Nobilis' in gardens. It does not seem to be the same as the pink-flowered species known today and illustrated in *Pelargoniums of southern Africa* since the flowers were much larger and the hypanthium again about three times the length of the pedicel. Another named *P. variegatum* (L.f.) Willd. is mentioned in several books and appears to have been very similar to the original large white-flowered one but with slightly broader stipules. This was illustrated by Cavanilles in 1787.

Pink-flowered plants could be confused with *P. patulum* or *P. tabulare* but Van der Walt[13] clearly sets out the differences. *P. patulum* is a decumbent plant with smaller flowers and petioled leaves on the flowering branches. *P. tabulare* is an erect plant but much smaller in height and flower than *P. grandiflorum*. However, the plant described in this paper as *P. grandiflorum* is considerably smaller than all the descriptions of the past. The proportions of the parts of the flower, the size of the stipules and the zoning on the leaf are not dissimilar to a plant illustrated by Jacquin and described by others up to the time of Knuth as *P. saniculaefolium* Willd. (syn. *P. cortusaefolium* Jacq.). *P. fuscatum* Jacq. was smaller and might be equivalent to *P. tabulare*[14]

of today. Jacquin had obviously not had the opportunity to see the white-flowered plant of Andrews which was introduced a few years later.

Very recently, a plant with unzoned leaves and large white flowers has been found growing wild in South Africa which may be similar to the original *P. grandiflorum*. Further observations will be needed to discover whether or not there is a complete range of characteristics in wild popu-lations, from the large white-flowered plants with unzoned leaves of the botanists of two hundred years ago, to the somewhat smaller pink-flowered plants with zoned leaves described in 1991. Perhaps this will prove that all, including *P. tabulare*, are variants of one quite variable species. Whatever the case, the discovery of a large, white-flowered unzoned plant is very exciting. *(Plate 13)*

P. laevigatum (Linnaeus filius) Willdenow, *Species plantarum*, **3:** 658 (1800)

MEANING: smooth, succulent, referring to leaves
SECTION: **Glaucophyllum**
ILLUSTRATION: Sweet, **3:** 294; *Pelargoniums of southern Africa*, **2:** 89.

Small very variable, often rather straggling plant, woody at base, usually glabrous, to about 50 cm (20 in) tall. Leaves about 3 × 2 cm (1 $^{1}/_{5}$ × $^{4}/_{5}$ in), variable but usually glaucous, slightly fleshy, divided into 3 linear or rounded segments, themselves 3-parted, each division rounded on under side, and with almost spine-like apices; petiole 1.5 cm ($^{3}/_{5}$ in), persistent; stipules awl-shaped, persistent on old stems. Flowers normally 1, white or pale pink, upper petals spathulate, marked with red, reflexed, *c.* 15 × 5 mm; lower petals smaller, unmarked; peduncle *c.* 3 cm (1 $^{1}/_{5}$ in); hypanthium *c.* 20 mm; pedicel 2–5 mm; fertile stamens 7.

RECOGNITION: rather straggling plant with glaucous, jointed, trifoliate leaves with linear pinnae and cream to pink flowers; similar to *P. fruticosum* and *P. ternatum*.

Although first collected by Thunberg around 1773, described in 1781, illustrated in several early books and known in gardens of Europe over two hundred years ago, it is not widely grown, despite being easy to cultivate. Even

Sweet in the text for his illustration in 1825 states that he thought it had already disappeared from cultivation but had managed to find a plant to illustrate from the nursery of Mr Lee. It grows wild in southern Cape Province and three subspecies have recently been recognized incorporating *P. oxyphyllum* DC. illustrated by Andrews, with trifoliate glabrous leaves and *P. diversifolium* Wendl. with hairy, sometimes unifoliate leaves.[15] *(Plate 14)*

P. lanceolatum (Cavanilles) Kern, *Hortus sempervirens*, **53**: 632 (1822)

MEANING: lanceolate, referring to leaf shape
SECTION: **Glaucophyllum**
ILLUSTRATION: *Botanical magazine*, **2**: 56; *Pelargoniums of southern Africa*, **1**: 18.

Upright branching plant, woody at base to 30 or 40 cm (12 or 16 in) tall. Leaves lanceolate, glaucous, apex acuminate, margins entire, 6–8 cm (2 $^2/_5$–3 $^1/_5$ in); petiole somewhat jointed at base, often persistent, 3–5 cm (1 $^1/_5$–2 in); stipules lanceolate, acuminate to 12 mm. Flowers 1–2, cream to pale yellow; upper petals marked with red, strongly reflexed, narrow oblanceolate, to 20 × 8 mm; lower petals slightly smaller, unmarked; peduncle 2–3 cm ($^4/_5$–1 $^1/_5$ in); hypanthium reddish, 3–4 cm (1 $^1/_5$–1 $^3/_5$ in); pedicel 1–2 mm; fertile stamens 7.

RECOGNITION: leaves lanceolate, undivided, flowers white to cream.

Aiton, in *Hortus Kewensis* (1789), says that this species was introduced by Kennedy and Lee 1775, but an interesting reference in the *Botanical Magazine* of 1787 notes that while Mr Lee, the owner of the Vineyard nurseries in Hammersmith, was looking at some dried specimens in the Banks herbarium of plants which had been recently received from the Cape of Good Hope, he found some viable seed which he then germinated and from which the illustration was made. So although he did not actually bring the plant back from South Africa, he did introduce it into cultivation. At first glance, this species in leaf, might not be recognized as a *Pelargonium* but the flower is quite typical and very similar to *P. laevigatum* in form. The branches are slightly brittle and may break easily but are easily re-rooted. It grows wild on rocky sites in south-western Cape Province. *(Plate 15)*

P. patulum Jacquin, N.J., *Collectanea*, **4**: 187 (1791)

MEANING: spreading referring to the habit
SECTION: **Glaucophyllum**
ILLUSTRATION: *Pelargoniums of southern Africa*, **2**: 113.

Trailing plant woody at base. Leaves variable, usually unzoned, glaucous, more or less rounded in outline, cordate at base, palmately 3-lobed, irregularly incised, *c.* 3 × 4 cm (1 $^1/_5$ × 1 $^3/_5$ in); petiole long to 9 cm (3 $^3/_5$ in); stipules large and semi-persistent. Flowers 2 or 3, pale pink; upper petals narrowly spathulate to 15 × 5 mm, with conspicuous red markings, reflexed; lower petals smaller with a short reddish line; peduncle to 4 cm (1 $^3/_5$ in) with small sessile leaves at base; hypanthium 10–25 mm; pedicel 5–25 mm; fertile stamens 6 or 7.

RECOGNITION: similar to *P. grandiflorum* and *P. tabulare* but with trailing decumbent habit, usually unzoned leaves and sessile leaf-like bracts on flowering stems.

In cultivation in Europe by 1791, this species was known in England a few years later. It grows wild in the south-western part of Cape Province, in partially shaded positions under rocks or shrubs. It has attractive foliage and flowers and in cultivation succeeds well in a hanging basket.

The variability of this species has given rise to its division into three varieties based on flower and foliage characteristics.[16] In var. *grandiflorum*, the flowers are larger and the leaf glaucous, whereas in var. *tenuilobum* the leaves are more deeply divided into five segments.

P. tabulare (Burman filius) sensu J.J.A. van der Walt, *Flowering plants of Africa*, **51**: 2035 (1991)

MEANING: from the Table Mountain
SECTION: **Glaucophyllum**
ILLUSTRATION: *Flowering Plants of Africa*, **51**: 2035 1991.

This species is similar in habit and appearance to *P. grandiflorum*, as it is known today, but smaller, rarely exceeding 50 cm (20 in) in height with the upper petals of the flower under 15 mm long. In size it is similar to *P. patulum* but has an erect not trailing habit, the leaves on the inflorescence are stalked and the pollen yellow, not orange. It is found in south-western Cape Province and was known in the middle of the eighteenth century. It was illustrated by Cavanilles in 1785 in Paris and as *P. fuscatum* by Jacquin from the gardens of Schönbrunn, near Vienna in 1792. It must not be confused with *P. tabulare* (L.) L'Hérit., as illustrated by L'Héritier, which is *P. elongatum* (Cav) Salisb., a totally unrelated plant with tiny cream or pale yellow flowers. This latter species has been listed in catalogues as *P. tabulare* and is far more widely grown. *(Plate 16)*

P. ternatum (Linnaeus filius) Jacquin, *Icones plantarum rariorum*, **3**: 11, t. 544 (1792)

MEANING: divided into three, referring to foliage
SECTION: **Glaucophyllum**
ILLUSTRATIONS: *Botanical magazine*, **12**: 413; Sweet, **2**: 165; *Pelargoniums of southern Africa*, **2**: 138.

Upright or spreading plant, woody at base to about 50 cm (20 in). Leaves green often with red margins, sometimes rough, slightly cupped, trifoliate, segments more or less triangular with tapered base and broad irregularly toothed apex, to 2 x 3 cm ($^4/_5$ x 1 $^1/_5$ in); petiole to 10 mm but upper leaves almost sessile; stipules narrow ovate; Flowers usually solitary, pale pink or white; upper petals narrow obovate, barely reflexed, with red-purple markings, c. 10 x 3 mm; lower petals smaller, unmarked; peduncle c. 1 cm ($^2/_5$ in) in axil of almost sessile leaf; hypanthium c. 15 mm; pedicel c. 3 mm; fertile stamens 5.

RECOGNITION: shrubby plant with trifoliate leaves and usually pale pink flowers borne singly; related to *P. laevigatum* and *P. fruticosum*.

This is one of the many species collected towards the end of the eighteenth century when several naturalists were exploring the southern tip of Africa. Kew's plant collector, F. Masson travelled on some journeys with Thunberg who sent a herbarium specimen in 1781 to Uppsala where it was first described. Kew received plants, or more likely, seeds from Masson in 1789 and it was also grown elsewhere in Europe. It does not have a wide natural distribution but is found in south-western Cape Province.

Hoarea (Sweet) de Candolle

Separated by Sweet as a distinct genus and named after a well-known contemporary pelargonium enthusiast, Sir Richard Colt Hoare, this is the largest section with possibly over 70 distinct species being recognized once it has been fully revised. In the wild, these species grow mainly in the arid inhospitable areas of winter rainfall of west and south-west Cape Province with a few extending into southern Cape Province. They are adapted to withstand the hot dry summers by dying down immediately after flowering and passing most of the year underground in a dormant state. This habit makes the plants difficult to find except during a limited period each season. The foliage appears after the winter rain but the flowers do not develop until the leaves have died down so that in the wild, it is rarely possible to examine the flowers and leaves at the same time.

As they are rarely concentrated in large colonies and may not even appear above ground every year if conditions are unfavourable, there may be many more species to be discovered. On the other hand, the morphology of several species is so variable that considerable investigation may be needed to discover whether or not the plant justifies the status of a unique species or whether it is simply a geographical variant of one already described. Many names have been given in the past which have later proved to be synonyms. To solve the many problems it would be necessary to gather them all together and grow them under standard conditions for comparison. However, this would not necessarily produce typical plants which could be equated with the same found in the wild.

All are geophytes with one or more tubers covered with papery sheaves which help to prevent the tuber drying out in summer. The remains of the petioles and stipules are usually clearly visible at the apex of the tuber and the leaves and flower stems emerge directly from the top. In some species the tubers may become quite large and have been used as food or for medicinal purposes by native people. The leaves are usually pinnately veined either simple or divided, sometimes into very fine segments. In the wild, they appear before the flowers but in cultivation this is not a reliable characteristic and dormancy is not always as prolonged as might be anticipated. The inflorescence, often branched, usually has many flowers and each flower may have either two, four or five petals and five, sometimes two, three or four fertile stamens. The flowers are very irregular and have a hypanthium exceeding the pedicel in length.

Several early botanists subdivided the section on the basis of the leaf shape but this may be a variable characteristic from one plant to another, one year to the next or even on one plant at one time. Lindley separated some species with the two lower stamens much longer than the rest into a section *Dimacria*, the classification which was adopted by Sweet for *P. pinnatum* in 1820. In 1824 Sweet placed the two petalled species into their own small section, *Seymouria*, named after the Honourable Mrs. Seymour of Woburn in Bedfordshire, who was described as 'an admirer of Geraniaceae and alpines'. In 1825, he also created a fourth genus, *Grenvillea*, for those species with four stamens. These characteristics are not considered sufficient to merit this division and all are now included in the one section *Hoarea*. This section may be distinguished from *Polyactium* by the covering of papery sheaves on the tubers, the more irregular flowers, and by the flowers which, at least in the wild, appear as the foliage is dying down. In cultivation, however, this last diagnostic characteristic is less useful. The basic chromosome for the majority of species is $x = 11$; most have 22 chromosomes.

Many were introduced in the early years of plant collecting in South Africa and their tuberous habit would have made them easy to maintain during the long journey back to Europe. For a short time during the early years of the nineteenth century, many species were grown and hybridized. These were illustrated in the literature of the time but perhaps because of the difficulties of cultivation and the long periods of dormancy, their popularity waned. Some species are grown in specialist collections but they are not the easiest pelargoniums to maintain and are very sensitive to excess water, especially during the dormant period. In their natural habitat, most grow in sandy soils. They appear to thrive when potted into a very free-draining compost of about half peat and half horticultural sand and watered only when in leaf, gradually reducing the water as the leaves die down and the flowers emerge. Bright light is important during the short growing season and a liquid fertilizer may be used at this time. They may be propagated by seed if it is available although the seedlings are very sensitive to overwatering. Alternatively the tubers may be separated and repotted after flowering.

The plants are slow-growing and the flowers do not last for very long before the whole plant dies down to remain underground for yet another year. However the flowers of many are very beautiful and worth the wait and some such as *P. incrassatum* with brilliant pink flowers, *P. rapaceum* with flowers shaped like those of a tiny pea flower or *P. appendiculatum* with its feathery foliage are sometimes available and not quite so temperamental. Of the many species described in the literature, only a selection have been included as examples of the variety that are grown by enthusiastic collectors. Very few are available commercially.

1	Petals 2	**asarifolium**
	Petals more than 2	2
2	Stipules conspicuous, over 2 cm ($^4/_5$ in) long	**appendiculatum**
	Stipules not exceeding 1.5 cm ($^3/_5$ in) long	3
3	Flowers resembling a tiny pea flower	**rapaceum**
	Flowers not as above	4
4	Stamens 2	**punctatum**
	Stamens usually 5	5
5	Flowers yellow or cream; leaves always simple	**oblongatum**
	Flowers not yellow or cream or if pale yellow, leaves variable or pinnate	6
6	Flowers bright magenta; scape unbranched	**incrassatum**
	Flowers never bright magenta; scape branched	7
7	Petals under 8 mm long	**auritum**
	Petals over 10 mm long	8
8	Petals more than 4 times as long as wide; leaves very variable	**longifolium**
	Petals less than 3 times as long as broad; leaves always pinnate	**pinnatum**

P. appendiculatum (Linnaeus filius) Willdenow, Species plantarum, **3:** 651 (1800)

MEANING: with appendages, referring to large stipules

SECTION: **Hoarea**

ILLUSTRATION: *Pelargoniums of southern Africa*, **3:** 6.

Plant covered with long hairs, with a large underground tuber covered with the remains of large stipules and leaf bases. Leaves grey-green, densely hairy, aromatic, oblong in outline, pinnate, each segment pinnately-divided and irregularly cut, to 10 × 5 cm (4 × 2 in); petiole to 10 cm (4 in); stipules to 3 × 1 cm (1 $^1/_5$ × $^2/_5$ in). Flowers yellow, up to 15 on long scape to 30 cm (12 in); upper petals spotted with red, narrow oblanceolate, notched at apex, *c.* 18 × 5 mm; lower petals unmarked, slightly narrower; hypanthium 5–10 cm (2–4 in); pedicel barely 1 mm; fertile stamens 5.

RECOGNITION: feathery grey-green leaves, large stipules and yellow flowers.

Collected by Thunberg in the late eighteenth century and illustrated by Cavanilles, it was sometimes included with *P. triste*, perhaps on account of the feathery foliage. It is not common in cultivation but being one of the larger, more vigorous species with a tall many flowered inflorescence, is worth growing.

P. asarifolium (Sweet) G. Don, *A general system of gardening and botany*, **1:** 731 (1831)

MEANING: leaf resembling that of *Asarum*

SECTION: **Hoarea**

ILLUSTRATION: Sweet, **3:** 206; *Pelargoniums of southern Africa*, **1:** 4.

Once included in the section Seymouria proposed by Sweet for this species with two petalled flowers, it is very rare in cultivation but included here as one example. The plants rarely exceed 15 cm (6 in) even in flower. The deep red or purple flowers are formed in pseudoumbels of about 10, on a branched scape. The two petals are linear in shape and 10 mm long. Arising directly from the underground tuber are rounded to heart-shaped leaves, hairy below, up to 4 cm (1 ³/₅ in) across. Found wild in south-western Cape Province, it was grown by Colvill's nursery in 1821.

P. auritum (Linnaeus filius) Willdenow, *Species plantarum*, 664 (1800)

MEANING: eared or auriculate
SECTION: **Hoarea**
ILLUSTRATION: Sweet **1**: 72 & **3**: 263; *Pelargoniums of southern Africa*, **1**: 21.

Tuberous plant to 10 cm (4 in) or more tall in flower. Leaves very variable from simple to bipinnatifid, with stiff hairs, the lamina to 12 cm (4 ⁴/₅ in) or more long; petiole exceeding lamina; stipules lanceolate. Flowers very dark purple or pink, in many flowered heads on a branching scape; upper petals linear to 10 x 3 mm, often less; lower petals very similar; hypanthium c. 10 mm; pedicel 1–2 mm; fertile stamens 5, equal in length and conspicuously exserted.

RECOGNITION: small flowers with narrow petals in many-flowered heads on branched scapes.

The narrow petals and exserted stamens give the flowers a star-like appearance. Two subspecies are recognized differing in their flower colour; subsp. *carneum* has white or pink flowers streaked with red while subsp. *auritum* has very dark purple petals with white bases. The first illustration of something close to the plant known today was of a very dark reddish-purple flowered plant, by Commelin in 1697 named *Geranium africanum, foliis plerumque auritis, floribus ex rubro purpurascentibus*, long before the days of the simple binomial name. It was collected, described and illustrated on many occasions after this, and as with many of this section with such polymorphous foliage, under many different names, some of which may eventually prove to have been correctly described as species. It may be found growing wild in south-western Cape Province.

P. incrassatum (Andrews) Sims, *Botanical magazine*, **20**: 761 (1804)

MEANING: thickened, referring to leaves or more likely to the tuber
SECTION: **Hoarea**
ILLUSTRATION: *Botanical magazine*, **20**: 761; Sweet, **3**: 262; *Flowering plants of Africa*, **29**: 1134; *Pelargoniums of southern Africa*, **2**: 78.

Plant with large fleshy underground tuber. Leaves with soft greyish-white hairs, narrow ovate in outline, deeply pinnately lobed, 3–6 x 2–5 cm (1 ¹/₅–2 ²/₅ x ⁴/₅–2 in); petiole equalling lamina; stipules narrow lanceolate, c. 15 mm. Flowers bright magenta, scape to 30 cm (12 in) tall bearing 20–40 flowers; upper 2 petals spathulate, paler at base, c. 20 x 8 mm; lower 3 petals spathulate with inrolled edge much smaller than upper c. 10 x 4 m; hypanthium to 4 cm (1 ³/₅ in); pedicel to 5 mm; fertile stamens 5.

RECOGNITION: bright magenta flowers in many-flowered inflorescence; leaves silky, silvery green.

Two names have been associated with this species since the time it was first collected for the Royal Botanic Gardens at Kew by Francis Masson in 1791. These and others with slightly paler flowers were given the illegitimate name, *P. roseum* which was first illustrated by Andrews in *The botanist's repository*. Later plants introduced into cultivation by Colvill's nursery in 1801 also illustrated by Andrews, were darker in colour and named *P. incrassatum* but the similarity between the two is only now recognized.

It should be noted that Sweet acknowledges that his own illustration is a copy of one by Mr Sydenham Edwards painted in 1802 and published in the *Botanical magazine*. He also adds that although the species seeds freely, and that Colvill in the past had houses filled with the plants, the plant had become very neglected by 1820 because of the interest in Cape Heaths. By illustrating the species again, he was trying to encourage its reintroduction into gardens. This species is found in the coastal regions of western Cape Province and Namaqualand where the rainfall does not exceed 12 cm (4 ⁴/₅ in) a year, but it is more common in cultivation nowadays than others of the section. Despite the limited flowering period, it is certainly worth the wait for the dramatic sight when the flowers appear quite early in the year.

P. longifolium (Burman filius) Jacquin, *Icones plantarum rariorum*, **3:** 9, t. 518 (1792)

MEANING: long leaves

SECTION: ***Hoarea***

ILLUSTRATION: *Flowering plants of South Africa*, **9:** 335; *Pelargoniums of southern Africa*, **1:** 25.

Hairy or glabrous plant reaching 25 cm (10 in) high in flower. Leaves very variable from simple ovate or lanceolate to lobed or bipinnately-divided into almost filiform segments; petiole often exceeding lamina; stipules filiform. Flowers white, yellow or pink, in branched inflorescence of up to 5 flowers per head; upper petals with deep red markings, narrow lanceolate to narrow obovate, 10-15 x 3-6 mm; lower petals slightly smaller; hypanthium *c.* 10 mm; pedicel 1-2 mm; fertile stamens 5.

RECOGNITION: small plant with very variable leaf formation and flower colour.

As many as twenty five names have been associated with this very polymorphous species but as in other variable species, some variants may be considered to be simply forms of a variable species and the names used as synonyms. Others may prove to be distinct and reinstated as individual species. Only the intensive investigations being carried out in South Africa will be able to resolve these difficulties of nomenclature. The first plant given this name in the illustration by Burman in 1759 basically had simple leaves, but some were pinnately incised.

Individual plants may bear simple and pinnate leaves at the same time or may have only one type, so it is not surprising that it is difficult to be certain which variations of the species were collected when or by whom. However several were illustrated by Jacquin in the late 1700s. Plants are found over a very wide area of the south-western parts of Cape Province in parts frequented by many early naturalists and explorers, so if they were travelling during the flowering season, specimens would almost certainly have been collected. *(Plate 17)*

P. oblongatum E. Meyer ex Harvey in *Flora capensis*, **1:** 263 (1860)

MEANING: oblong, referring to tuber

SECTION: ***Hoarea***

ILLUSTRATION: *Botanical magazine*, **98:** 5996; *Pelargoniums of southern Africa*, **1:** 29.

Hairy plant with a vertical oblong tuber exposed at apex. Leaves usually ovate or rounded with cordate base and large blunt teeth, 10–15 x 5–8 cm (4–6 x 2–3 $^1/_5$ in); petiole *c.* 4 cm (1 $^3/_5$ in); stipules lanceolate *c.* 2 cm ($^4/_5$ in) or more. Flowers large, pale yellow borne on branching scapes to 25 cm (10 in); upper petals with dark maroon feathered veins, obovate, *c.* 30 x 15 mm; lower petals unmarked, smaller; hypanthium 4–5 cm (1 $^3/_5$–2 in); pedicel *c.* 1 mm; fertile stamens 5, long, curved upwards.

RECOGNITION: large pale yellow flowers, simple leaves and oblong tuber.

The native habitat of this species is the northern part of Namaqualand which might explain why, despite the very large spectacular flowers, it was not collected until 1814 by Burchell, much later than many other pelargoniums and not named until many years after this. It is no more difficult to cultivate than others of the section as long as it is kept quite dry during the dormant period.

P. pinnatum (Linnaeus) L'Héritier, Aiton, *Hortus Kewensis*, **2:** 417 (1789)

MEANING: pinnate, referring to leaves

SECTION: ***Hoarea***

ILLUSTRATION: Sweet, **1:** 46; *Pelargoniums of southern Africa*, **1:** 34.

Stemless plant with relatively large tuber. Leaves bluish green, with silky hairs, oblong in outline, pinnate with elliptic pinnae each *c.* 10 x 5 mm; petiole 3–6 cm (1 $^1/_5$–2 $^2/_5$ in); stipules linear, joined to petiole for half their length. Flowers clear pink but pale yellow and white forms may be found in the wild; inflorescence of several branches each up to 15 flowered; upper petals with red feathering, obovate, *c.* 20 x 8 mm; lower petals unmarked, slightly smaller; hypanthium 20–25 mm; pedicel 1–2 mm; fertile stamens 5.

RECOGNITION: small plant with pinnate leaves and usually pink flowers.

If grown in a light soil and watered carefully, this species will produce a mass of pink flowers each spring preceded by attractive bluish-green leaves; it appears to set seed quite easily in cultivation. The nomenclature has been quite confused as it is not certain whether it is variable in both flower and leaf morphology or if the different forms are in fact unique species. Very similar if not identical plants are illustrated by Commelin in 1703 where he notes that seeds had been received from the Cape of Good Hope, but without mention of a date. It is not surprising that it was collected on several other occasions as it is very striking in flower and would be found on the normal routes across south-western Cape Province. Many names have been associated with *P. pinnatum* but plants with narrower petals and narrow leaflets are now referred to as *P. viciifolium* DC. (syn *G. pinnatum* Andr.) and those with undulate petals, the upper curved, are named *P. trifoliolatum* (Eck. & Zeyh) E.M. Marais (syn. *P. astragalifolium* (Andr.) Loudon).[17] *(Plate 18)*

P. punctatum (Andrews) Willdenow, *Species plantarum*, **3**: 645 (1800)

MEANING: spotted, referring to petals
SECTION: **Hoarea**
ILLUSTRATION: *Pelargoniums of southern Africa*, **2**: 119; Flowering plants of Africa **53**: 2108.

This species superficially shows some similarity to *P. oblongatum* in habit and leaf shape and its yellow flowers. It may be distinguished in flower by the very narrow pale yellow petals barely 2 mm wide, the upper spotted red much longer than the lower, and the stamens reduced to 2 in number. The flowers have a rather unpleasant smell in full bloom but are numerous and very striking in appearance. The leaves are rather small, broad ovate, to 2 x 2.5 cm ($^4/_5$ x 1 in), with a truncate base and petiole to 1.5 cm ($^3/_5$ in). The tuber has an irregular and more globose shape than *P. oblongatum*. According to Aiton, it was collected in 1794 by Thomas Johnes. It has also been grown at the gardens of the Royal Horticultural Society, Wisley, for several years but is almost unknown except in specialist collections.

P. rapaceum (Linnaeus) L'Héritier, Aiton, *Hortus Kewensis*, **2**: 418 (1789)

MEANING: turnip-shaped, referring to tuber
SECTION: **Hoarea**
ILLUSTRATION: *Botanical magazine*, **44**: 1877; Sweet, **1**: 18 & **2**: 135; *Pelargoniums of southern Africa*, **1**: 39.

Plant with one large and several smaller tubers. Leaves hairy, linear, pinnate, the pinnae further divided into linear segments forming whorls along the rachis, to 10 x 2 cm (4 x $^4/_5$ in); petiole almost equal to lamina; stipules narrow triangular. Flowers similar in shape to a legume flower, pink or yellow, on scape to 12 cm (4 $^4/_5$ in) or more, bearing several heads of 10 or more flowers; upper petals held erect together, with red markings, obovate *c.* 15 x 4 mm; lower petals held together like a keel enclosing the stamens, unmarked, broader than upper; hypanthium 10–15 mm; pedicel 2–3 mm; fertile stamens 5.

RECOGNITION: flowers shaped like a pea flower, leaves finely divided.

There has been considerable confusion in the use of this specific name in older literature, especially in North America, and even in catalogues of some nurseries today. The totally unrelated cultivar 'Mrs Kingsbury', with scented foliage and large cerise-coloured flowers, is incorrectly listed under *P. rapaceum*. 'Mrs Kingsbury' is a shrub-like plant to several feet in height, normally classified with the Uniques.

The variability of the flower colour and markings in the true species has resulted in the existence of many descriptions of species or varieties. Those names found most frequently are var. *selinum* with pale pink flowers, var. *luteum* with bright yellow flowers and darker spots at the base of the upper petals, and var. *corydalifolium* with pale yellow flowers, the upper petals marked red at the base.

It was one of the earliest of the section to be collected and early references of Commelin and Hermann at the end of the seventeenth century, illustrate and describe this species with its typical flower shape and swollen root. Later authors including Linnaeus in his first edition of *Species Plantarum* in 1753, confused the name with *P. myrrhifolium*. In the wild, it has a very wide distribution through Cape Province. The tubers were used both as a food and for their medicinal properties. *(Plate 19)*

Isopetalum (Sweet) de Candolle

Sweet assigned **P. cotyledonis**, the only species in this section, to its own genus and gave it the name *Isopetalum* to indicate the similar shape and size of the petals, but the genus was reduced to a section by de Candolle in 1824. It resembles plants of the *Otidia* section in its swollen knobbly stems and branched inflorescence and at times has been included within this section. The unmarked white, more or less radially symmetrical flowers have five or six fertile stamens and an exceedingly short hypanthium. The plants have 22 chromosomes.

P. cotyledonis (Linnaeus) L'Héritier, Aiton, *Hortus Kewensis*, **2:** 428 (1789)

MEANING: from leaves resembling *Cotyledonis*
SECTION: *Isopetalum*
ILLUSTRATION: Sweet, **2:** 126; *Pelargoniums of southern Africa*, **3:** 33.

Plant of about 30 cm (12 in), with short, slightly branched, stout, succulent stems covered with scaly bark and rough with persistent stipules, becoming brownish with age. Leaves clustered at ends of branches often falling in dry season, rounded with cordate base or almost peltate, 2–5 cm (⁴/₅–2 in) across, leathery, rather glossy dark green, above with conspicuous impressed veins, densely grey hairy beneath; petiole to 8 cm (3 ¹/₅ in); stipules narrow triangular. Flowers white, regular, 1 cm (²/₅ in) across, on a branched inflorescence; peduncle to 5 cm (2 in), bearing up to 15 white flowers; all petals elliptic, unmarked, *c.* 12 × 6 mm; hypanthium 1 mm; pedicel to 8 mm; fertile stamens 5, occasionally 6.

RECOGNITION: succulent stems and more or less regular unmarked white flowers.

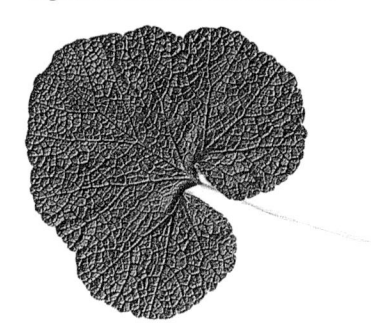

This species is endemic to the island of St Helena in the southern Atlantic where it grows on rocky cliffs often exposed to salt sea spray. Unfortunately like many species of the native flora, it has become very rare in its native country and wild goats appear to have been one of the major causes of its decline. However, the programme of work being carried out on the island, led by the Conservation unit from Kew, is halting this decline and the species is no longer considered to be seriously endangered. Grown as a curiosity and known as 'old man live forever', it has not lost its appeal since being introduced to Kew in 1765 by John Bush. Separated geographically from the rest of the genus and adapted to the conditions of this isolated island, it is quite unique and easily distinguished from all other *Pelargonium* species.

P. cotyledonis may be grown in a free-draining medium of equal quantities of horticultural silver sand and soilless compost but is very sensitive to overwatering, especially in winter. It can be propagated by cuttings of the younger growths. However, if possible it is better to grow from seed as the plant is very slow-growing and even one cutting can spoil a plant which may take several years to recover its shape. *(Plate 20)*

Jenkinsonia (Sweet) Harvey

In his first volume of *Geraniaceae* in 1820–1822, Robert Sweet proposed a new genus, *Jenkinsonia*, which he named after Robert H. Jenkinson, one of the well-known collectors of members of the family *Geraniaceae* in the early nineteenth century. As well as *P. tetragonum* and *P. praemorsum*, he included one or two species showing similar morphological characteristics, which are now classified in the closely related section *Myrrhidium*. The genus was demoted to a section of *Pelargonium* by de Candolle. Old references especially in German, also refer to a genus *Chorisma* which was used by Ecklon and Zeyher for *P. tetragonum*. All grow in regions with very hot dry summers and winter rainfall, conditions to which they have adapted in different ways.

The characteristic features of this section include the very zygomorphic flowers with two exceedingly large upper petals and two or three very much smaller lower ones, seven prominent fertile stamens which are often bent upwards almost at a right angle and a long hypanthium usually at least twice as long as the pedicel and often much more. The leaves are palmately lobed and at least in the South African species, are rather small and soon dropped to conserve water as the dry season begins. The habits of the plants are extremely diverse. Of the three South African species, *P. tetragonum* is unique with its angled, green, succulent, jointed and rather brittle stems and either a tufted, pendant or a scrambling habit depending on the situation in which it is growing. The other two South African species also have a somewhat jointed habit, but more woody stems and persistent stipules which in the typical form of *P. antidysentericum* become spiny. The inflorescence of these three species bears only one or two large flowers each with four petals.

Two other species are the exceptions in the *Pelargonium* world as not only are they hardy in many parts of British Isles, they are also native to the Middle East in the northern hemisphere. These two have a herbaceous habit, swollen underground roots for survival during adverse conditions and larger more persistent leaves than those of their South African relatives. The inflorescences are also distinct, bearing many flowers instead of one or two but the similar flower structure puts them into this section.

P. boranense from East Africa with a swollen stem and orange-red flowers is also rather an anomaly but is included in this section at present. There are however suggestions that *Jenkinsonia* is an artificial section which might be better split into two or more smaller ones and, together with other characteristics, the variable chromosome numbers found among the species would support this idea. For *P. antidysentericum* and *P. praemorsum* the basic chromosome number is 9 but for *P. tetragonum* and *P. boranense* $x = 11$. For *P. endlicherianum* and *P. quercetorum* the diploid number is 34.

Cuttings usually root readily and seed is produced by most species. Conditions including bright light, a well-drained, gritty compost and careful watering to emulate the natural habitat, are best for cultivated plants. The leaves often drop and the plants become dormant for part of the year so during this time watering is rarely necessary. *P. endlicherianum* and *P. quercetorum* are the only two pelargoniums to grow quite happily out of doors in all but the very coldest parts of the British Isles as long as the plants are kept in a very freely-draining soil.

1. Stems 3 or 4 angled, always succulent **tetragonum**
   Stems not 3 or 4 angled 2
2. Plants with very short stems and leaves arising from apex 5
   Plants without this habit 3
3. Flowers scarlet **boranense**
   Flowers not scarlet 4
4. Plants with swollen tuber **antidysentericum**
   Plants without a swollen tuber **praemorsum**
5. Leaves shallow-lobed with blunt teeth; petals twice as long as wide **endlicherianum**
   Leaves deeply lobed with sharp teeth; petals three times as long as wide **quercetorum**

P. antidysentericum (Ecklon and Zeyher) Kosteletzky, *Allgemeine Medizinisch-pharmazeutische Flora*, **5:** 1896 (1836)

MEANING: name for medicinal use
SECTION: **Jenkinsonia**
ILLUSTRATION: *Pelargoniums of southern Africa*, **2:** 6.

Erect plant to 1 m (40 in) or trailing, with a very large partly underground tuber and slender branches becoming woody with age. Leaves on short branches, appearing as clusters, rounded, five-lobed, toothed, *c.* 15 mm x 15 mm, aromatic; petiole longer than leaves; stipules persistent forming hard curved spines in type species, otherwise membranous and deciduous. Flowers 2–3, purplish-pink or white veined with purple; peduncle to 20 mm; upper petals to 20 x 8 mm narrow spathulate; lower to 12 x 5 mm; hypanthium to 30 mm; pedicel to 10 mm; fertile stamens 7, prominently exserted; filaments slightly curved upwards.

RECOGNITION: related to *P. praemorsum* but distinguished by presence of a tuber and less showy flowers.

This unusual species is found growing in very arid regions and in cultivation needs bright light and a gritty soil with very little water, especially in winter. Cuttings may root but propagation of the tuber is more successful unless fertile seed is produced. In the wild the leaves are shed during the summer and reappear after flowering in autumn but in cultivation with less extreme temperatures, the leaves may remain on the plant. Recently, three subspecies have been described.[18] In subsp. *antidysentericum* the stipules become persistent recurved spines whereas those of subsp. *inerme* Scheltema, are deciduous and membranous. The third subspecies, *zonale* Scheltema, has membranous stipules, a zoned leaf, paler flowers and a more trailing habit. The underground tubers which may reach nearly two metres in diameter in the wild, were used by the Hottentots as a medicine in cases of dysentery and anaemia. First described in 1835, it was known to be in cultivation a few years later.

P. boranense Friis and Gilbert, *Flowering Plants of Africa*, **43:** 1705 (1976)

MEANING: named after the place where it was first found, Borana Awraja in Ethiopia

SECTION: **Jenkinsonia**

ILLUSTRATION: *Flowering Plants of Africa*, 1705.

Perennial plant with a short upright swollen, succulent stem bearing the remains of persistent leaf bases. Leaves borne towards the top of stem, to 15 cm (6 in) across, broader than long, three lobed, the lateral lobes sometimes further divided, the central lobe stalked, or further divided and the leaflets coarsely toothed; petiole red tinged, to 20 mm with enlarged base; stipules, deciduous, minute. Flowers 2–5, scarlet or orange-red; peduncle to 50 cm (20 in); upper petals 2, with darker markings, broad spathulate, to 4 x 2 cm (1 $^3/_5$ x $^4/_5$ in); lower 2, slightly smaller, unmarked; hypanthium to 2 cm ($^4/_5$ in); pedicel 2–3 cm ($^4/_5$–1 $^1/_5$ in); fertile stamens 7.

RECOGNITION: large, bright red, 4-petalled flowers.

This species was first found growing wild in grassland in the highlands of Ethiopia in 1972 and a few plants are cultivated by enthusiasts. Plants are less tolerant of lower temperatures than most other species and become dormant for much of year when water should be withheld. Water freely when in growth but use a very well-drained compost and avoid a humid atmosphere.

P. endlicherianum Fenzl, *Pugillus plantarum novarum Syriae et Touse occidentalis* (1842)

MEANING: named after Stephan Endlicher, Director of the Botanic Garden in Vienna until 1849, and predecessor of Fenzl, the authority for the name

SECTION: **Jenkinsonia**

ILLUSTRATION: *Botanical magazine*, **82:** 4946; *Kew Magazine*, **10:** 227.

Herbaceous perennial, with a very short stem above ground producing a rosette of leaves from an underground rhizome. Basal leaves orbicular, with 5 shallow, crenate lobes, to 6 cm (2 $^2/_5$ in) across; petiole to 7 cm (2 $^4/_5$ in); stipules triangular, *c.* 10 mm. Flowers 5–15, slightly scented bright purplish-pink; peduncle to over 20 cm (8 in) bearing a few small leaves; upper petals recurved, spathulate, to 30 x 15 mm; lower petals 3, minute or absent; hypanthium to 20 mm long; pedicels *c.* 5 mm long; fertile stamens 7, exserted; filaments curved upwards.

RECOGNITION: rosette-like habit with bright purplish-pink flowers with exceptionally large upper petals but minute lower ones; related to *P. quercetorum*.

One of the few species which have an award from the Royal Horticultural Society, *P. endlicherianum* received an Award of Merit in 1901, exhibited as a rock garden plant. Exceptionally for this genus, it may be grown out of doors in the British Isles, in a sunny dry situation especially if sheltered from excess rain. However, it is easier to care for in an alpine house where the watering, especially in winter, is more easily regulated. In the wild, it is widespread in open rocky situations in Turkey where it was first discovered in the Taurus mountains by a traveller named Kotschy. It was sent to the gardens of the Emperor of Austria in Vienna and first flowered at Kew in 1856, grown from seed sent from the Botanic Garden in Copenhagen. There was obviously a good exchange of plants at this time between different countries and it must also have been known in Belgium as an article in *Belgique Horticole* of the same year suggests that there would soon be hybrids available. However, either no one succeeded or tried, as there are no records to be found. *(Plate 21)*

P. praemorsum (Andrews) Dietrich, *Lexicon der Gartnerei und Botanik*, **7**: 48 (1807)

MEANING: 'bitten off', referring to the blunt ended leaflets

SECTION: **Jenkinsonia**

ILLUSTRATION: *Botanical magazine*, **15**: 547; Sweet, **1**: 79; *Pelargoniums of southern Africa*, **1**: 35.

Plant to 30 cm (12 in) or so with branching rather thin stems giving a zig-zag appearance. Leaves soon deciduous, almost round, *c.* 5–30 mm across, divided into five coarsely toothed, wedge-shaped leaflets, sometimes clustered at the nodes; petiole usually about equal to lamina; stipules persistent, rigid. Flowers 1–2, cream or pink to purple; peduncle up to 10 cm (4 in); upper petals obovate, to 35 × 30 mm, veined with deep red or purple; lower 2 or 3, faintly marked, half size of upper; hypanthium to 4 cm (1 $^3/_5$ in); pedicel *c.* 10 mm; fertile stamens 7, exserted, filaments strongly curved upwards.

RECOGNITION: very large 'butterfly-like' flowers with woody zig-zag stems; related to *P. antidysentericum*.

Two subspecies have been recently recognized; one, subsp. *praemorsum* has cream flowers and pale green leaves, and the second, subsp. *speciosum* Scheltema[18] has pink or purple flowers and dark green, densely hairy leaves. According to the *Botanical magazine*, this plant was first seen in flower in England in 1801, raised from seed introduced in 1798 by Mr Quarrell collecting new plants for Colvill's Nursery in the Kings Road, Chelsea. On the other hand, *Hortus Kewensis* claimed that F. Masson introduced it earlier in 1793. Perhaps the Kew plants did not survive to flower.

Whatever the true history, it is not too difficult to grow and produces spectacular cream or purplish-coloured flowers in early summer. In its native habitat in Namaqualand, it usually grows in the shelter of rocks in a region with very little winter rain. In the wild the leaves are dropped to conserve water in the very hot dry summers although this does not necessarily happen consistently in cultivation.

P. quercetorum Agnew, *Kew bulletin*, **21**: 225-227 (1967)

MEANING: from oak woods (*Quercus libani*)

SECTION: **Jenkinsonia**

ILLUSTRATION: *Pelargoniums of southern Africa*, **3**: 115; *Kew magazine*, **10**: plate 228.

This species is very similar to *P. endlicherianum*, and will thrive under similar conditions but in flower is distinguished by the much narrower upper petals to 30 mm × 10 mm. The thickened roots produce short aerial stems covered with persistent triangular stipules and the larger, coarsely toothed, deeply lobed leaves are rough to the touch. It grows wild in the rich soils of moist oak woods in northern Iraq and SE Turkey but because of its recent discovery is still uncommon in cultivation.

P. tetragonum (Linnaeus f.) L'Héritier, Aiton, *Hortus Kewensis*, **2**: 427 (1789)

MEANING: stem 4-angled
SECTION: **Jenkinsonia**
ILLUSTRATION: *Botanical magazine*, **4**: 136; Sweet, **1**: 99; *Pelargoniums of southern Africa*, **1**: 45.

A sprawling or upright tufted plant with green, 3 or 4-angled, succulent stems; long internodes each up to 20 cm (8 in) long, jointed at nodes. Leaves few, soon deciduous, somewhat succulent, rounded or reniform, shallow-lobed and crenate, 2–3 x 4 cm ($^4/_5$–1 $^1/_5$ x 1 $^3/_5$ in), usually with some hairs, dark green, often marked with a darker zone; petiole *c.* 3 cm (1 $^1/_5$ in); stipules small, deciduous. Flowers large, cream or pale pink, 1–2 at the ends of the branches; upper petals spathulate, veined dark red, to 40 x 15 mm; lower petals 2, half size of upper; hypanthium to 6 cm (1 $^2/_5$ in); pedicel very short; fertile stamens 7, strongly exserted; filaments sharply bent upwards.

RECOGNITION: angled green succulent stems with few leaves.

It was collected by F. Masson on his second expedition around the Cape Peninsula and introduced to Kew in 1774. Growing wild in dry rocky ground from the west coast to Grahamstown, it may be seen scrambling through shrubs or cascading down cliffs or rocky outcrops. The succulent green stems enable the plant to survive even when the leaves have fallen in the dry season and the jointed stems, though brittle are easily rooted if accidentally broken. Several references are made in the literature to plants with variegated leaves, such as one with a cream margin as shown in illustrations of both Andrews and Sweet. However, Sweet does mention that this variegation is lost when the plants are grown luxuriantly and no consistently variegated plants are found in gardens today so most, if not all, variegations are probably due to cultural conditions. *(Plate 22)*

Ligularia Harvey

The name for this section was used as a genus by Ecklon and Zeyher but Harvey in *Flora Capensis* made it into a section which was retained and enlarged by Knuth in 1912. It is, however, one of the most diverse sections and it is sensible to split it into more than one, as has been proposed by several taxonomists working on all aspects the genus.

Species are included which show similarities to almost all other sections. Some have tubers, some are semi-succulent, some spiny, herbaceous or woody. The flowers may be red, pink, white or cream with five or seven stamens. The foliage may be aromatic or not, simple or divided and species may be found in almost all habitats and geographical areas of southern Africa. A study of the chromosome numbers and sizes as well as the results of attempted hybridization between species of different sections was undertaken recently to try to rationalize the classification of the species at present included in this section *Ligularia*.[19] In most cases the results were confirmed by morphological similarities and also their natural geographical locations.

Some species such as *P. abrotanifolium*, have been moved into the new section *Reniformia*, two into *Cortusina* and one, *P. setulosum*, into *Campylia*. *P. barklyi*, *P. articulatum* and *P. aridum* appear close to *Ciconium* with a similar morphology and with the same chromosome number (x = 9). Hybrids are known with species of the latter section. *P. alpinum* has chromosomes showing a similarity to those of section *Pelargonium* but it is very dissimilar in morphological characteristics. The remaining species can be separated into smaller groups of related species. *P. dolomiticum*, *P. griseum* and *P. tragacanthoides* (x = 10) have very distinct flowers with four petals of which the upper two have curious claws rolled longitudinally to form a tube. All grow towards the eastern side of southern Africa although *P. dolomiticum* has a very much wider natural distribution. *P. divisifolium*, *P. plurisectum*, *P. tenuicaule* and *P. trifidum* (x = 10) may be found on the western side but are less obviously similar in appearance. *P. grandicalcaratum*, *P. karooicum*, *P. otaviense* and *P. spinosum* (x = 10) grow in very arid habitats of Namibia and the north and western regions of Cape Province and have a morphology adapted to survive summer drought. *P. exhibens*, *P. mollicomum* and *P. worcesterae* (x = 11) which grow in eastern areas of south Africa, appear a little different with morphological and geographical similarities to the section *Reniformia* but with a different chromosome number. The remaining members of the existing section with a basic chromosome number x = 11 include *P. crassipes*, *P. fulgidum*, *P. hirtum*, *P. hystrix*, *P. oreophilum*, *P. pulchellum*, *P. sericifolium* and *P. stipulaceum*. All have fleshy stems, often covered with the remains of stipules, sometimes becoming quite hard. Most are adapted to survive in regions of quite low rainfall on the far western side of South Africa.

There is therefore the probability that at least three new sections will be created. It is interesting to note that many of the species described below and included for the present in this section, are of more recent introduction, and were not known to the early botanists at the height of the interest in the new Cape introductions at the end of the eighteenth and beginning of the nineteenth centuries.

1. Plants more or less stemless — 2
   Plants with definite stems, though sometimes short — 5
2. Flowers white with red markings; leaves pinnately lobed — **pulchellum**
   Flowers and leaves not as above — 3
3. Plants with very large stipules — **stipulaceum**
   Plant without very large stipules; leaves often zoned — 4
4. Plant with a tuber — **barklyi**
   Plant without a tuber but root sometimes swollen — **articulatum**
5. Leaves toothed or lobed but never divided repeatedly or deeply — 6
   Leaves deeply or repeatedly divided or compound — 12
6. Petals held together so flower does not appear to completely open — 7
   Flowers with spreading petals — 8
7. Leaf base cordate — **otaviense**
   Leaf base cuneate or rounded — **grandicalcaratum**
8. Stems with persistent spiny petioles — **spinosum**
   Stems not spiny — 9
9. Flowers pink or purplish-pink — 10
   Flowers cream to white; foliage aromatic — 11
10. Upper petals over 2 cm ($^4/_5$ in) long with conspicuous dark purple blotch — **alpinum**
    Upper petals under 2 cm ($^4/_5$ in) long with lines but without conspicuous dark purple blotch — **rodneyanum**
11. Leaves aromatic; stems with long internodes and a straggly habit — **worcesterae**
    Leaves pineapple-scented; stems very short giving a compact habit — **mollicomum**
12. Petals 4, upper 2 with claws forming a cylinder — 13
    Petals 5, upper petals more or less flat — 15
13. Flowers dark pink — **griseum**
    Flowers white, cream or pale pink — 14

14	Lamina elongate in outline; petioles rough	**tragacanthoides**
	Lamina ovate in outline; petioles not rough	**dolomiticum**
15	Flowers scarlet	**fulgidum**
	Flowers not scarlet	16
16	Plant with thickened, almost fleshy or semi-succulent stems densely covered with remains of persistent stipules or petioles	17
	Plant woody or herbaceous; stems often with remains of petioles but not thickened, nor almost fleshy or semi-succulent	21
17	Flowers cream to white	**hystrix**
	Flowers pink	18
18	Flowers bright vivid pink, over 30 mm across	**sericifolium**
	Flowers pink, under 25 mm across	19
19	Petals broadly obovate	**hirtum**
	Petals oblong or narrow obovate	20
20	Base of petiole expanded, becoming hard and bending downwards with age; stipules joined to petiole for most of their length	**crassipes**
	Petiole bases not expanded and only base of stipule joined to petiole	**oreophilum**
21	Plant mainly herbaceous, sometimes woody at base	22
	Plant with distinctly woody stems	26
22	Flowers pink	**divisifolium**
	Flowers not pink	23
23	Stamens 5	**exhibens**
	Stamens 7	24
24	Petals broad with large purple blotch	**tenuicaule**
	Petals narrow without large blotch	25
25	Leaves divided into linear segments, without strong scent	**aridum**
	Leaves trifid or trifoliate but not divided into linear segments, with strong scent	**trifidum**
26	Plant erect; hypanthium over 2 cm ($4/5$ in) long	**plurisectum**
	Plant often pendant; hypanthium under 1.5 cm ($3/5$ in) long	**karooicum**

P. alpinum Ecklon & Zeyher, *Enumeratio plantarum africae australis extratropicae*, **1**: 80 (1835)

MEANING: mountainous, referring to habitat

SECTION: *Ligularia*

ILLUSTRATION: *Pelargoniums of southern Africa*, **3**: 1.

Low-growing plant becoming straggly as flowering stems elongate. Leaves often with thin dark zone towards margin, cordate, dentate and sometimes with shallow lobes, 3–5 × 2–4 cm (1 $1/5$–2 × $4/5$–1 $3/5$ in); petiole *c.* 5 cm (2 in); stipules ovate, over 1 cm ($2/5$ in). Flowers pink to orange-pink, large, with rather flat appearance; inflorescence unbranched, usually 2-flowered; peduncle 3–8 cm (1 $1/5$–3 $1/5$ in); upper petals with conspicuous dark red blotch and veins, broad obovate, *c.* 25 × 15 mm; lower petals unmarked, slightly smaller; hypanthium *c.* 2–4 cm ($4/5$–1 $3/5$ in); pedicel under 1 cm ($2/5$ in); fertile stamens 7, 2 very short.

RECOGNITION: low-growing plant with large salmon-pink flowers.

A form seen in cultivation with a conspicuously zoned leaf makes an attractive and fairly compact foliage plant before it becomes more untidy on flowering but the zone is not sharply defined at all seasons of the year. It grows wild in high mountainous regions of south-western Cape Province and in the wild is hardy enough to withstand snowfalls in winter. It was first collected in about 1830 but is not often seen in cultivation. *(Plate 24)*

P. aridum R.A. Dyer, *Kew bulletin*, 445 (1932)

MEANING: thought to refer to appearance of dormant plants

SECTION: *Ligularia*

ILLUSTRATION: *Pelargoniums of southern Africa*, **3**: 11.

Tufted plant with long conspicuous hairs and rather short stems covered with the remains of persistent petioles. Leaves orbicular in outline but divided into linear segments with inrolled margins, to 7 cm (2 ^{4}/$_5$ in) across; petiole over twice as long as lamina, lower part persistent; stipules narrow triangular, semi-persistent. Flowers white or pale yellow, up to 5 on peduncle to about 15 cm (6 in); upper petals narrow obovate, sometimes with a few short red lines at base, to 25 x 5 mm; hypanthium to 5 cm (2 in); pedicel 2–3 mm; fertile stamens 7, 2 very short.

RECOGNITION: tufted plant with very short internodes; flowers cream or white resembling *P. barklyi*.

This species is found over a wide area of eastern Cape Province and northwards through Orange Free State. It was first collected by Ecklon and Zeyher in the early part of the nineteenth century.

P. articulatum (Cavanilles) Willdenow, *Species plantarum*, **3**: 356 (1800)

MEANING: articulated, referring to rhizome which alternates between thick and thin portions giving appearance

SECTION: **Ligularia**

ILLUSTRATION: *Pelargoniums of southern Africa*, **2**: 11.

Plant with very short stem above ground but an underground rhizome which is alternately thick and thin. Leaves more or less orbicular with deeply cordate base, margin toothed and lobed, usually with a narrow dark zone towards margin, softly-hairy, *c.* 5 cm (2 in) or more across; petiole to 10 cm (4 in); stipules triangular to 15 mm. Flowers cream to white, large; inflorescence few flowered on long rather stout, peduncle; upper petals obovate, veined with purple, 30 x 15 mm; lower petals much narrower and slightly shorter; hypanthium to 6 cm (2 ^{2}/$_5$ in) or more; pedicel very short; fertile stamens 7.

RECOGNITION: large cream-coloured flowers and orbicular, dark green, often zoned leaves. Differs from *P. barklyi* in absence of tuber.

This species has been used in Australia with zonal cultivars to produce a range of interesting hybrids. The similar chromosome number of $x = 9$ and the hybrids resulting in the crosses mentioned, indicate a close relationship with section *Ciconium* and justifies the suggestion that, together with *P. barklyi* and *P. aridum*, it should be removed to *Ciconium* from *Ligularia*. It is thought that Thunberg was the first to collect this species in the early 1770s from the western part of Cape Province. It is more tolerant of moist soils than many others.

P. barklyi Scott Elliot, *Journal of botany*, **29**: 68 (1891)

MEANING: named for Sir Henry Barkly, Governor of Cape of Good Hope in 1870s

SECTION: **Ligularia**

ILLUSTRATION: *Pelargoniums of southern Africa*, **3**: 17.

Geophytic plant with short aerial stem and covered with long, conspicuous hairs. Leaves dark green usually with narrow dark purplish zone near margin, often purplish on underside, orbicular, with deeply impressed veins, margin dentate, *c.* 5 cm (2 in) across; petiole to 15 cm (6 in) or more; stipules triangular to 10 mm. Flowers very pale yellow, up to 5 on each long peduncle to 15 cm (6 in) or more; upper petals narrow obovate, *c.* 25 x 7 mm; lower petals slightly smaller; hypanthium *c.* 5 cm (2 in); pedicel 3–4 mm; fertile stamens 7, 2 very short.

RECOGNITION: a tuberous plant with dark green, usually zoned leaves and relatively large, cream coloured flowers.

The large tuber of this species was the reason why it was for many years included first in the section *Polyactium* and later in *Hoarea*. It is now thought to show considerable affinity with *Ciconium*. It grows in north-western Cape Province in areas of low rainfall where the tuber enables it to survive the hot dry summers. It is not difficult to cultivate as long as it is kept dry while dormant, and is interesting for the unusual foliage though it has only been recognized for just over one hundred years.

P. crassipes Harvey, *Flora Capensis*, **1**: 281 (1860)

MEANING: from latin 'crassus' meaning thick, referring to stem

SECTION: **Ligularia**

ILLUSTRATION: *Pelargoniums of southern Africa*, **1**: 10.

Small plant to about 30 cm (12 in) tall with a thick main stem, very short internodes and covered with deflexed, hardened petiole bases. Leaves hairy, oblong in shape, bipinnately-divided into narrow segments, 3–7 cm (1 $^{1}/_{5}$– 2 $^{4}/_{5}$ in) long; petiole about equal to lamina, broadened to base; stipules joined to petiole for most of their length. Flowers pink lined with darker red, 5–10 on branching peduncle bearing a few small leaves; upper petals obovate, c. 10 x 4 mm; lower petals slightly smaller; hypanthium 5–7 mm; pedicel more or less equal to hypanthium; fertile stamens 7.

RECOGNITION: small plant with thick unbranched stem covered with hardened, flattened, recurved petioles; flower pink. Somewhat similar in habit to *P. hystrix* and *P. oreophilum* but with flattened petiole bases.

This species was first collected by Masson but not named until 1860 when Harvey included it in *Flora Capensis*. It grows wild in Namaqualand in the very arid regions with a little rain in winter.

P. divisifolium Vorster, *South African journal of botany*, **53**: 71 (1987)

MEANING: divided leaves

SECTION: **Ligularia**

ILLUSTRATION: *Pelargoniums of southern Africa*, **3**: 48.

Somewhat scrambling, glabrous plant, slightly woody at base. Leaves ovate in outline but 2–3 times pinnately-divided into linear segments each rarely exceeding 1 mm wide, about 5–6 cm (2–2 $^{2}/_{5}$ in) long but may be more when growing vigorously; petiole c. 5 cm (2 in); stipules very narrow, to over 10 mm. Flowers pink, large, 3–5 on peduncle usually over 5 cm (2 in); upper petals with distinct dark red marks and veining, broad obovate, to 20 x 10 mm; lower petals unmarked, narrow obovate, c. 15 x 5 mm; hypanthium 15–30 mm; pedicel about equal to hypanthium; fertile stamens 5.

RECOGNITION: plant with finely divided leaves, long stipules and large pink flowers.

Although this species was clearly illustrated by Knuth in 1912 as *P. artemisifolium*, there has been so much confusion over the plant, its name and its relationship with *P. abrotanifolium* that for taxonomic reasons it was renamed *P. divisifolium* in 1987. However, as the species was probably not in cultivation until it was added to specialist collections relatively recently, this should cause no confusion. It does not appear difficult to grow and is worth cultivating for the large attractive pink flowers. It may be found wild in a small area of south-western Cape Province. *(Plate 25)*

P. dolomiticum Knuth, *Sitzungsberichte der Königliche preussischen Akademie Wissenschaften*, **11**: 877 (1906)

MEANING:, referring to dolomite rock formation where species was first found

SECTION: **Ligularia**

ILLUSTRATION: *Flowering plants of South Africa*, **20**: 780; *Pelargoniums of southern Africa*, **2**: 54.

Perennial plant which dies back in winter and with thickened tuberous rootstock. Leaves bipinnate, lamina ovate in outline, to 10 x 6 cm (4 x 2 $^{2}/_{5}$ in); petiole often equalling or exceeding lamina; stipules ovate. Flowers white or pale pink, in branched inflorescence, each pseudo-umbel of 3–5 flowers; upper petals strongly reflexed, with small red markings, claws inrolled, narrow oblong, to 20 x 3 mm; lower petals 2, shorter but broader than upper; hypanthium c. 5 mm; pedicel 2–4 mm; fertile stamens 7.

RECOGNITION: flowers with four petals, the upper two with claws rolled to form a tube, distinguished from *P. griseum* by white or pale pink flowers and from *P. tragacanthoides* by tuber-ous root, smooth petioles and broader leaf.

P. dolomiticum is a variable plant with a wide distribution in southern Africa and plants may be found in regions of very low or quite high rainfall falling in summer or winter. In the past, plants from the eastern part of its range were known as *P. bechuanicum*. It is one of the several species collected in the early part of the nineteenth century which were not fully recognized until nearly a hundred years later and are still very rarely cultivated.

P. exhibens Vorster, *South African journal of botany*, **52:** 481 (1986)

MEANING: used to describe stamens which are clearly exserted

SECTION: **Ligularia**

ILLUSTRATION: *Pelargoniums of southern Africa*, **3:** 54.

This recently described species is somewhat similar to *P. trifidum* but can be immediately recognized by the foliage which does not have a strong unpleasant scent. It forms a plant of about 30 cm (12 in) or more in height, rarely branched and with stems covered with the bases of semi-persistent petioles. The leaves with a thickened texture, are trilobed to 6 cm (2 $^2/_5$ in) across with petioles to 7 cm (2 $^4/_5$ in) and membranous stipules to 4 mm long. Compared to *P. trifidum*, the white to cream-coloured flowers are slightly smaller, the hypanthium is considerably shorter and there are 5 fertile stamens. It grows in eastern Cape Province and although it may have been collected earlier was unrecognized as a distinct species and only named in 1986.

P. fulgidum (Linnaeus) L'Héritier

MEANING: from latin 'fulgidus' referring to brightly coloured flowers

SECTION: **Ligularia**

ILLUSTRATION: Sweet, **1:** 69; *Flowering plants of Africa*, **37:** 1475; *Pelargoniums of southern Africa*, **I:** 16.

Plant spreading or scrambling, to about 1 m (40 in), softly-hairy with somewhat succulent stems. Leaves with a silver sheen, oblong, pinnately lobed, often with two almost free lobes at base, the lobes irregularly toothed, to 10 x 7 cm (4 x 2 $^4/_5$ in); petiole *c.* 5 cm (2 in); stipules broad ovate. Flowers scarlet; inflorescence branched with 4–9 flowered heads; upper petals veined darker red, broad oblong, strongly reflexed, to 20 x 10 mm, veined darker; lower petals slightly narrower; hypanthium often dark brownish-red, very conspicuous and swollen at base, to 18 mm; pedicel *c.* 8 mm; fertile stamens 7.

RECOGNITION: bright red flowers and leaves with a silvery sheen.

This is one of the older species to be introduced into Holland in the early eighteenth century and from there sent to Italy and England. It is one of the few members of the genus which appears able to hybridize with species of several other sections such as *Polyactium*, *Glaucophyllum* and *Pelargonium*, and is involved in the parentage of many of the red-flowered cultivars such as *P.* 'Ardens' and 'Scarlet Unique'. It grows along the west side of South Africa, near to the coast. Like many red-flowered plants with long nectary tubes, it is pollinated by birds in the wild. Occasionally pink-flowered plants may be found. It is not difficult to grow but many examples seen in cultivation do not appear to be the true species and do not bear fertile pollen, so attempts at reproducing the older hybrids from these will be unsuccessful.

P. grandicalcaratum Knuth, *Repertorium specierum novarum regni vegetabilis*, **15:** 135 (1918)

MEANING: large prominent hypanthium
SECTION: **Ligularia**
ILLUSTRATION: *Pelargoniums of southern Africa*, **3:** 60.

This species is very rarely seen in cultivation and was described for the first time in 1918. It grows in the arid regions along the western side of Cape Province and Namibia. It forms a small deciduous shrub to about 50 cm (20 in) in height, with slightly fleshy, obovate, somewhat peppery-scented, leaves, toothed at the apex, about 10 mm or more long. The flowers are quite distinct with a very conspicuous, thickened hypanthium to about 15 mm long and a much thinner shorter pedicel, which like the calyx is tinged red. The flowers appear closed as the two small upper petals bend slightly backwards while the much larger lower three form a tube-like structure enclosing the seven fertile stamens. The petals are almost oblong in shape, white to cream, veined and flushed with dark red, the lower over 1 cm ($^2/_5$ in) long. The only other species with an apparently closed flower is *P. otaviense* with larger flowers and leaves with a cordate not cuneate base.

P. griseum Knuth, *Das Pflanzenreich 4*, **129:** 439 (1912)

MEANING: grey, referring to grey-green leaves
SECTION: **Ligularia**
ILLUSTRATION: *Pelargoniums of southern Africa*, **2:** 67.

This species, not collected until 1861 or named until 1912, may be found wild in Cape Province but is still very rarely seen in collections except in its native land. The unusually shaped flowers have four petals, the upper two with claws rolled longitudinally to form a tube. It may be distinguished from the closely related *P. dolomiticum* and *P. tragacanthoides* by its dark pink flowers. It forms a small plant, woody at the base, the stems covered with the remains of persistent petioles and with grey-green hairy leaves pinnately-divided into narrow segments.

P. hirtum (Burman f.) Jacquin, *Icones plantarum rariorum*, **3:** 10, t. 536 (1792)

MEANING: hairy
SECTION: **Ligularia**
ILLUSTRATION: Sweet, **2:** 113; *Pelargoniums of southern Africa*, **1:** 22.

Low-growing hairy plant with erect or spreading, thickened branches brownish with age and covered with persistent petioles. Leaves feathery, carrot-like, faintly aromatic, to about 5.5 x 2 cm (2 $^1/_5$ x $^4/_5$ in); petiole to 4 cm (1 $^3/_5$ in); stipules lanceolate, acute attached to petiole for at least half their length. Flowers bright to purplish-pink, almost regular in appearance with spreading petals; inflorescence leafy with several clusters of 2–6 flowers, on peduncle *c.* 6 cm (2 $^2/_5$ in); upper petals obovate, with darker spot at base, 11 x 6 mm; lower petals slightly narrower without darker spots; hypanthium *c.* 5 mm; pedicel to 4 mm; fertile stamens 7.

RECOGNITION: small plant with soft feathery leaves and pink flowers with an almost regular appearance. Similar to *P. oreophilum* but the persistent stipules do not become hard.

P. hirtum has been cultivated in European gardens for over two hundred years and is not difficult to grow or to propagate by cuttings or from seed. It grows wild in the Cape Peninsula mainly on sandy soils.

P. hystrix Harvey, *Flora Capensis*, **1**: 280 (1860)

MEANING: taken from 'hystrichos', greek for porcupine referring to spiny persistent stipules

SECTION: **Ligularia**

ILLUSTRATION: *Pelargoniums of southern Africa*, **2**: 77; *Flowering plants of Africa* **53**: 2109.

Similar to *P. crassipes* with short thick stems covered with the remains of hard persistent stipules giving it a spiny appearance. When flowering, it reaches about 20 cm (8 in) or so in height and the flowers are pale cream-coloured or white with narrow oblong petals, feathered with dark red markings, to 15 mm long. The leaves are ovate, pinnately-divided into narrow segments to about 3 cm (1 $^1/_5$ in) long on a petiole also to 3 cm (1 $^1/_5$ in). The stipules are triangular to almost 10 mm becoming very hard after the leaves fall. It grows in arid inland areas of south-western Cape Province where the leaves are lost for the whole of the hot summers. First collected by F. Masson during the latter part of the eighteenth century, it is not easy to maintain unless treated as a succulent with great care taken with the watering and the free-drainage of the compost.

P. karooicum Compton & Barnes, *Transactions of the royal society of South Africa*, **19**: 295 (1931)

MEANING: refers to karoo, the vegetation type where the plant was first found

SECTION: **Ligularia**

ILLUSTRATION: *Pelargoniums of southern Africa*, **3**: 78.

Deciduous plant woody at base with drooping or spreading branches, semi-succulent when young but becoming woody after a season, bearing hard remains of petiole bases. Leaves thick and fleshy, ovate in outline, usually divided into linear segments but sometimes only lobed, to 2 cm ($^4/_5$ in) long; petiole 5–8 mm; stipules minute. Flowers pink, white or pale yellow, usually 2 on each peduncle to about 2 cm ($^4/_5$ in); upper petals veined with dark red, very narrow spathulate, notched at apex, reflexed, to 20 x 5 mm; lower petals unmarked and wide spreading; hypanthium *c.* 10 mm; pedicel *c.* 2 mm; fertile stamens 5.

RECOGNITION: shrubby plant with long, rarely branched, often pendant stems without leaves for long periods of the year.

(actual size)

This species from south-western Cape Province, grows in inland areas with very little rain and in cultivation needs a very well-drained soil, bright light and almost no water when the leaves have fallen. It is able to survive long periods without attention and when not in flower or leaf, is very insignificant which may explain why it was not collected until the end of the last century. It is however an interesting curiosity which is not difficult to cultivate and any branches which are easily broken, may be rooted without difficulty. The name should not be confused with the cultivar 'Karooense' found in gardens, which is similar in flower and foliage to 'Graveolens'.

P. mollicomum Fourcade, *Transactions of the royal society of South Africa*, **21**: 92 (1932)

MEANING: softly hairy, referring to leaves

SECTION: **Ligularia**

For many years this species has been associated with *P. odoratissimum* but although it shows a superficial similarity, the foliage has a distinct pineapple scent which makes it a very useful plant in a scented or species collection. It is a more substantial plant than *P. odoratissimum* with a short thick stem, persistent, pale, narrow triangular stipules and longer hairs on the stems and leaves. The light green leaves often have a darker zone towards the base and are more or less rounded with a truncate or cuneate base, lobed with crenate margins, to 4 cm (1 $^3/_5$ in) or more across. The flowers are creamy-white and larger than those of *P. odoratissimum*, borne on rather thick peduncles. The fruits are quite noticeably swollen resembling those of some members of the section *Ciconium*. It is closely related to *P. worcesterae* which has a more straggly habit, and is also found in eastern Cape Province.

P. oreophilum Schlechter, *Botanische Jahrbücher*, **27:** 151 (1900)

MEANING: from Greek meaning 'loving mountains' referring to its natural habitat
SECTION: **Ligularia**
ILLUSTRATION: *Pelargoniums of southern Africa*, **2:** 102.

This species shows some similarity to *P. crassipes* and *P. hystrix* with short thickened stems covered with the hardened remains of persistent stipules and petioles. It differs from *P. crassipes* in the much larger flowers with narrow obovate, pink petals to 15 mm long with dark red marks or blotches at the base. The leaves are ovate in outline to 2 cm (⁴/₅ in) long, and pinnately-divided into linear divisions on petiole to about 2 cm (⁴/₅ in). The petioles are semi-persistent but the narrow triangular stipules about 5 mm, harden and remain on the stems for several seasons. Plants of this species should be grown in very free-draining soils and bright light. They may be found wild in mountainous regions in a small area of south-western Cape Province.

P. otaviense Knuth, *Das Pflanzenreich* 4, **129:** 439 (1912)

MEANING: from Otavi Mountains in Namibia
SECTION: **Ligularia**
ILLUSTRATION: *Pelargoniums of southern Africa*, **2:** 107.

Erect, deciduous woody plant to about 1 m (40 in) with stems covered with persistent petioles. Leaves cordate with large teeth, aromatic with lemony scent, to 4 cm (1 ³/₅ in) across; petiole persistent, to 4 cm (1 ³/₅ in); stipules narrow triangular. Flowers white or pinkish, up to 6 on peduncle to 3 cm (1 ¹/₅ in) or more; petals similar, narrow obovate, to 15 x 5 mm, held upright so that flower does not appear to open; hypanthium slightly thickened, to 15 mm; pedicel *c.* 5 mm; fertile stamens 7.

RECOGNITION: shrub with white or pinkish-coloured flowers which appear closed, separated from *P. grandicalcaratum* by larger flowers and reniform leaves and from *P. spinosum* by the flower shape and less spine-like petiole remains.

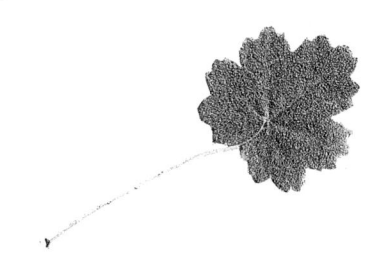

This is another relatively newly described species which was first collected by Professor Dinter in 1908 from the Otavi Mountains from which the species derives its name. It grows over a wide area of the western part of Namibia in full sun to altitudes of over 2,000 m (6,500 ft) but often in regions with very low rainfall. It is not common in cultivation but its natural habitat would indicate that it should be given bright light and very free-draining soils.

P. plurisectum Salter, *Journal of South African botany*, **8:** 279 (1942)

MEANING: repeated divisions, referring to leaves
SECTION: **Ligularia**
ILLUSTRATION: *Pelargoniums of southern Africa*, **2:** 114.

Initially this species could be mistaken for *P. abrotanifolium* in the section *Reniformia*. It has a similar twiggy appearance with thin woody stems covered with the remains of persistent petioles and tiny finely divided, blue-green leaves, rarely over 1 cm (²/₅ in) in length. It differs however in several features, especially the lack of the characteristic scent. Stipules are present in this species, to 3 mm long, joined to the petiole base. The flowers of both species have a long hypanthium and very short pedicel and only one or two are formed on each peduncle, but those of *P. plurisectum* are slightly smaller and the petals a creamish-pink colour veined with pink. It is not commonly seen in cultivation but is found wild in a small area of south-west Cape Province and was known originally from a collection made by Zeyher in the early part of the nineteenth century.

P. pulchellum Sims, Curtis, *Botanical magazine*, **15**: 524 (1801)

MEANING: from latin meaning small and beautiful
SECTION: **Ligularia**
ILLUSTRATION: *Botanical magazine*, **15**: 524; Sweet 1: 31; *Flowering plants of Africa*, **42**: 1669; *Pelargoniums of southern Africa*, **1**: 36.

Hairy plant with a rosette habit and very short, rarely branched, thick semi-succulent stem and thickened root, reaching about 15 cm (6 in) in flower. Leaves softly-hairy, oblong, pinnately lobed to pinnatifid, c. 6 x 4 cm (2 $^2/_5$–1 $^3/_5$ in); petiole 2–4 cm ($^4/_5$–1 $^3/_5$ in); stipules large, lanceolate, fused to base of petiole. Flowers white in a many flowered umbel in branching inflorescence; upper petals with faint red lines at base, sometimes red blotched, obovate to 20 x 8 mm; lower petals with prominent red blotch, slightly smaller; hypanthium to 4 cm (1 $^3/_5$ in); pedicel c. 5 mm; fertile stamens 7.

RECOGNITION: plant with short stem, pinnately lobed leaves; flowers white, the lower three petals with red blotches.

This is one of the most striking species of the genus which if grown in a very free-draining soil in bright light, is not too difficult to grow and will flower early in the season. It differs from many as the marking on the petals are on the lower not the upper. Records note that it was introduced to Kew by F. Masson in 1795 but was almost certainly collected at other times. Illustrations by William Paterson, who collected plants on an expedition to north-western Cape Province with Gordon when the Orange River was first discovered about twenty years earlier, depict a very similar plant. It grows wild in Namaqualand in full sun but very little rain.

P. rodneyanum Mitchell ex Lindley, Mitchell, *Three expeditions into the interior of Eastern Australia*, **2**: 143 (1838)

MEANING: presumably named for the British Admiral Rodney who died in 1792
SECTION: **Ligularia**

Plant with brown tuberous roots and short aerial stems. Leaves ovate, base sometimes cordate, margin crenate, or obscurely lobed, to 4 x 4 cm (1 $^3/_5$ x 1 $^3/_5$ in); petiole thin, to twice length of lamina; stipules ovate, 4–7 mm. Flowers dark rose- or purplish-pink, inflorescence branched with about 5 flowers in each umbel; upper petals with darker pink markings obovate, c. 15 x 7 mm; lower petals narrower; hypanthium to c. 5 mm; pedicel c. 1 cm ($^2/_5$ in); fertile stamens 7.

RECOGNITION: tuberous, short-stemmed plant with quite large deep rose pink flowers.

This may be found in south-eastern and South Australia, often growing among rocks, and was first collected on the expedition of Major Mitchell in 1836 to southern Australia. Once included in section *Polyactium* on account of the tubers, it is included here in the section *Ligularia* until the classification of this diverse section has been resolved. It has recently been introduced into a few specialist collections and forms a very floriferous plant which is not difficult to grow and looks especially attractive in a hanging basket. It may be propagated from the small tubers as well as by seed. *(Plate 26)*

P. sericifolium J.J.A. van der Walt, *Journal of South African botany*, **46**: 284 (1980)
MEANING: silk-like, referring to hairs on leaves
SECTION: **Ligularia**
ILLUSTRATION: *Pelargoniums of southern Africa*, **2**: 125.

Low-growing, branching plant with slightly thickened stems covered with the remains of stipules and petioles which do not become excessively hard. Leaves covered with short silky hairs giving a silvery-green colour, palmatisect, c. 15 × 10 mm; petiole persistent, about equal in length to lamina; stipules persistent, linear to 5 mm, fused to petiole for most of their length. Flowers large, bright purplish-pink with dark markings in centre of each petal, 1 or 2 on peduncle of 3–5 cm (1 $^1/_5$–2 in); upper petals broad obovate, c. 20 × 15 mm; lower petals c. 15 × 5 mm; hypanthium to 5 cm (2 in) or more; pedicel very short; fertile stamens 7.

RECOGNITION: small plant with vivid magenta pink flowers and silvery-green leaves.

(actual size)

In flower colour, this species could be mistaken for *P. magenteum* but differs in almost every other characteristic. It was first found in 1838 and grows in a limited area in north-western Cape Province with very low rainfall. It is rare in cultivation but the silvery foliage and brilliantly coloured flowers make it a very attractive addition to a collection as long as bright light and a free-draining compost can be provided.

P. spinosum Willdenow, *Species plantarum*, **3**: 681 (1800)
MEANING: spine, referring to persistent petioles covering stems
SECTION: **Ligularia**
ILLUSTRATION: *Pelargoniums of southern Africa*, **2**: 132.

Deciduous plant, woody at base, the thick, pale-coloured stems covered with the hardened remains of persistent petioles which form long spines with persistent thorn-like stipules at their base. Leaves slightly fleshy with a rather pungent scent, usually cordate with coarse teeth, in two sizes: the smaller leaves are borne on very short deciduous petioles and the larger of 2–3 cm ($^4/_5$–1 $^1/_5$ in) across on persistent petioles to over 10 cm (4 in); stipules to 2 cm ($^4/_5$ in). Flowers white or pale pink, large, up to 10 on peduncle to about 5 cm (2 in); upper petals veined purple, narrow obovate, to 30 × 10 mm; lower petals wide spreading, faintly marked, slightly smaller; hypanthium c. 2 cm ($^4/_5$ in); pedicel usually exceeding hypanthium; fertile stamens 7, 2 very short.

RECOGNITION: vegetatively similar to *P. otaviense* but with spines from persistent petioles and thorns from persistent stipules, as well as larger open pink flowers.

This is one of the species described and illustrated by William Paterson during his exploration in 1778. F. Masson later introduced it into cultivation nearly twenty years later and it is still grown by a few enthusiasts. It does not flower reliably in cultivation but plants can be grown in very bright light and a very well-drained soil similar to that of its native habitat in the arid lands of southern Namibia and north-western Cape Province. *(Plate 27)*

P. stipulaceum (Linnaeus filius) Willdenow, *Species plantarum*, **3**: 655 (1800)
MEANING: with stipules
SECTION: **Ligularia**
ILLUSTRATION: Sweet, **2**: 148; *Pelargoniums of southern Africa*, **2**: 132.

Almost hairless plant with a thickened woody rootstock, reaching about 30 cm (12 in) in flower, with a rosette-like habit and a very short, rarely branched, thick semi-succulent stem bearing pale, conspicuous, persistent stipules.

Leaves light green with fruity scent, cordate, toothed, pinnately lobed often with two almost free lobes at base, to 5 x 4 cm (2 x 1 ³/₅ in); petiole usually much longer than lamina; stipules ovate to over 1 cm (²/₅ in). Flowers pale yellow, in few-flowered umbels on branching leafy stems; upper petals *c.* 20 x 7 mm; lower petals slightly smaller, with red lines at base; hypanthium 4–6 cm (1 ³/₅–2 ²/₅ in); pedicel minute; fertile stamens 7.

RECOGNITION: plant with short thickened stem, similar to *P. pulchellum* but with pale yellow flowers and less divided leaves.

This species has been known since Thunberg sent material to Sweden where it was named by the son of Linnaeus in 1781. It grows in an area of western and south-western Cape Province in regions where the rainfall may be as little as 10 cm (4 in) a year so in cultivation it needs very well-drained soils with very careful attention to the watering regime. Like *P. pulchellum*, and other species from Namaqualand, it flowers very early in the season. *(Plate 28)*

P. tenuicaule Knuth, *Repertorium specierum novarum regni vegetabilis*, **45:** 63 (1938)

MEANING: long, thin branches
SECTION: **Ligularia**
ILLUSTRATION: *Pelargoniums of southern Africa*, 3: 138.

Plant with small tubers and straggling spreading stems. Leaves slightly fleshy, pale green, but more or less orbicular in outline, with 5 shallow toothed lobes, to 4 cm (1 ³/₅ in) or more across; petiole usually much longer than lamina; stipules 3–4 mm. Flowers opening cream but becoming white, large with a rounded, shallow-cupped shape, in pairs on long branching inflorescence; upper petals with very prominent dark purple blotch, spathulate, *c.* 20 x 10 mm; lower petals unmarked, diamond-shaped, sometimes slightly larger than upper; hypanthium up to 2 cm (⁴/₅ in); pedicel 2–3 mm; fertile stamens 7.

RECOGNITION: straggling plant with large rounded flowers opening cream, becoming white with a large dark purple blotch on each upper petal.

Despite its attractive flowers, this plant can become rather untidy but is ideal for growing in a hanging basket. It may be propagated from the small tubers but must be given a very well-drained soil, limited water and bright light. In the wild, it is known in southern Namibia and northern Namaqualand where it may be found in rocky areas.

P. tragacanthoides Burchell, *Travels in the interior of southern Africa*, **2:** 98 (1824)

MEANING: referring to scent of foliage from gum tragacanth
SECTION: **Ligularia**
ILLUSTRATION: *Pelargoniums of southern Africa*, 3: 144.

Erect, rather stiff, strongly scented plant, woody at base. Leaves hairy, pinnately to bipinnately-divided giving a feathery appearance, narrow ovate to almost lanceolate in outline, to 8 x 2 cm (3 ¹/₅ x ⁴/₅ in); petiole harsh or rough to touch, *c.* 3 cm (1 ¹/₅ in); stipules triangular to 5 mm. Flowers white, 2–4 on peduncle of 4–6 cm (1 ³/₅–2 ²/₅ in); upper petals sharply reflexed, veined with red, apex notched, claws rolled, *c.* 20 x 6 mm; lower petals 2, smaller, without rolled claws, notched apex or markings; hypanthium *c.* 5 mm; pedicel 1–2 mm; fertile stamens 7, conspicuously exserted.

RECOGNITION: flowers with four petals, the upper two with claws rolled to form a tube, distinguished from *P. griseum* by white flowers and from *P. dolomiticum* by non-tuberous root, rough petioles and narrower leaf.

The leaves of this plant have been used medicinally in its native land where it may be found in a wide area from eastern Cape Province to

Namibia. It is thought to have been known for over two hundred years but early illustrations of *P. ramosissimum* show plants similar in habit but which may not be the same because the flowers have five petals. The species described here was collected by Burchell in 1813 but although easy to grow, is not widely cultivated, perhaps because of the strong, and not very pleasant scent of the foliage.

P. trifidum Jacquin, *Plantarum rariorum horti caesarei Schoenbrunnensis*, **2**: 5, t. 134 (1797)

MEANING: 3-parted, referring to leaves

SECTION: **Ligularia**

illustration: *Pelargoniums of southern Africa*, **1**: 15.

Plant woody at base but with scrambling herbaceous stems. Leaves trifid to trifoliate, bright green, slightly thickened and fleshy, with strong unpleasant scent when touched, 2–4 cm ($^4/_5$–1 $^3/_5$ in) across; petiole 2–4 cm ($^4/_5$–1 $^3/_5$ in); stipules ovate, 3–4 mm. Flowers large, cream to almost white, 3–5 on long peduncle; upper petals with conspicuous red veining, narrow obovate, *c.* 25 x 7 mm; lower petals smaller, unmarked; hypanthium over 2 cm ($^4/_5$ in); pedicel 1–2 mm; fertile stamens 7.

RECOGNITION: rather straggly plant with strongly-scented foliage and relatively large creamy-white flowers.

Although many people find the foliage of this species rather unpleasantly scented when touched, the plant can be grown in a hanging basket and make a very attractive feature when covered with the relatively large, freely produced flowers. Some forms with larger more deeply divided leaves and slightly larger flowers unfortunately seem also to have a much more pungent scent resembling rotting fish. It is easy to grow and seeds freely and may be found wild in arid regions from the south-western to the eastern regions of Cape Province. It was first introduced into cultivation at Kew by Masson under two names in 1790 and 1794, neither of which is valid today. *(Plate 29)*

P. worcesterae Knuth, *Botanische Jahrbücher*, **40**: 75 (1907)

MEANING: presumably from the area where found in the wild

SECTION: **Ligularia**

This species has been confused with *P. dichondrifolium* which forms a tufted plant with very short internodes and a stem covered with long persistent petioles. However, *P. worcesterae* is quite distinct with rounded tubers formed underground, and long internodes giving a rather straggly habit. The brown stems have broad-triangular, very pale green stipules which are persistent. The leaves are aromatic, almost sticky to touch, reniform, the margins crenate or shallow-lobed, somewhat grey-green, to 4 cm (1 $^3/_5$ in) across with a petiole of 2–5 cm ($^4/_5$–2 in) or more. The white flowers have narrow petals *c.* 15 x 5 mm and 5 fertile stamens. The hypanthium is over 10 mm but the pedicel very short. The flowers are similar in shape to *P. exhibens* but it is easily distinguished as the latter has much more deeply divided non aromatic foliage. *P. worcesterae* may be found in eastern Cape Province in areas with low rainfall which falls in summer.

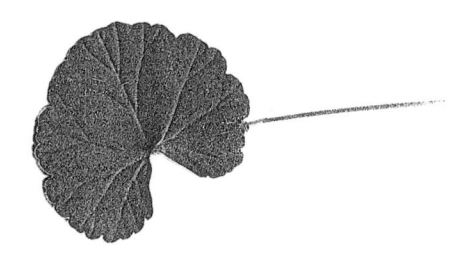

Myrrhidium de Candolle

In contrast to many other sections, these plants have no obvious morphological adaptation to the dry seasons. However, in their native country, many die back after flowering and pass the extreme conditions as seed ready to germinate as soon as the rain falls again. Frequently found as sprawling roadside weeds in the winter rainfall areas of south-western Cape Province, some extend further east and north into the summer rainfall regions. Many are pioneer plants which quickly colonize disturbed ground. Most are herbaceous with pinnately-divided leaves and rather long straggly stems sometimes woody at the base. The flowering stems become very straggly and elongated, with smaller leaves along their length, and long peduncles holding the flowers above the foliage.

After flowering, these long stems die back or may be pruned back to the woody basal stems but sometimes the whole plant will die back so it is advisable to retain the freely produced seed. The inflorescence is few-flowered and each flower is very irregular in shape with two very large upper petals, usually with long claws, but three or sometimes only two, small lower ones. The sepals have a membranous texture and are conspicuously veined and ribbed. The hypanthium may be quite long and the fertile stamens five or seven. The fruits enlarge quite considerably as they ripen and the mericarps seem large in proportion to the size of the plant. The very zygomorphic flowers with the upper petals considerably larger than the lower, demonstrates the similarity of this section to *Jenkinsonia* and some species were at one time included within the former genus *Jenkinsonia*. The basic chromosome number is x = 11 and all have 22 large chromosomes except *P. caucalifolium* which has 44.

Plants of this genus are often very variable depending on their habitat and the species are sometimes difficult to separate. This has resulted in the creation of more section and generic names than for any other section. The resemblance of the flowers of *P. myrrhifolium* and *P. longicaule* to those of *P. tetragonum* encouraged Sweet to include them together in the genus, *Jenkinsonia* but de Candolle in 1824 created the section *Myrrhidium* for all those known at the time. In 1831, George Don reversed the decision reinstating the genus *Jenkinsonia*, but Ecklon and Zeyher in 1835 changed the classification yet again and raised the section *Myrrhidium* once more to generic level. Another genus *Geraniospermum*, was described by Knutze at the end of the century but fortunately this name did not last long. Many of the species are variable and have been known under a number of names but a recent revision has managed to resolve most of the earlier problems.[20] The very rarely used synonyms have been deliberately omitted from the index of this book but may be found in the revision.

In cultivation they are easy to grow although rather short-lived. They do not necessarily die down completely after flowering if watered regularly and the flowering shoots cut back. Seed is large and freely produced, germinating more readily than that of many other species of the genus which are notoriously slow or erratic. Cuttings are also successful. All succeed best in a light sandy soil. Many are not spectacular as garden plants but several flower over a long season, and their trailing habit lends them to cultivation in a hanging basket where their large flowers are shown to advantage.

1 Upper petals over 2 cm ($^4/_5$ in) long   2
   Upper petals under 2 cm ($^4/_5$ in) long   4
2 Peduncle with 1, rarely 2 flowers
                                    caucalifolium
   Peduncle normally with over
   3 flowers   3
3 Flower buds large and conspicuous
   enclosed in large bracts   **suburbanum**
   Flowers buds not large and
   conspicuous   **longicaule**
4 Hypanthium over 1 cm ($^2/_5$ in);
   stamens prominently exserted   **multicaule**
   Hypanthium under 1 cm ($^2/_5$ in);
   stamens not prominently exserted   5
5 Leaves grey-green, lobed   **candicans**
   Leaves green, more deeply
   divided   **myrrhifolium**

P. candicans Sprengel, *Systema vegetabilum*, **3**: 57 (1826)

MEANING: white, referring to silvery sheen on leaves

SECTION: **Myrrhidium**

ILLUSTRATION: *Pelargoniums of southern Africa*, **1**: 6.

This species shows some resemblance to *P. myrrhifolium* var. *myrrhifolium* with rather small pinkish-coloured flowers. It is however distinguished by the leaves which are a grey almost silvery-green, often with a darker blotch in the centre and with two lobes at the base. The flowers always have four petals and five fertile stamens. It is found wild over a large area of the coastal strip of south-western and southern Cape Province, where it grows in both mountains and at lower altitudes. It was first recognized and introduced in 1802 but for many years was thought to come from the Canary Islands. It is a pretty but not a showy plant and very short-lived, often dying after a single season.

P. caucalifolium Jacquin, *Icones plantarum rariorum*, **3**: 10, t. 529 (1792)

MEANING: leaves similar to those of *Caucalis* sp., a small member of Umbelliferae found in Europe

SECTION: **Myrrhidium**

ILLUSTRATION: subsp. *caucalifolium*, *Pelargoniums of southern Africa*, **3**: 18; subsp. *convolvulifolium*, *Pelargoniums of southern Africa*, **3**: 23.

Superficially this species might be mistaken for *P. longicaule* but in *P. caucalifolium* the flowers rarely have more than four petals and there is usually only one flower, occasionally two on each flower stem. The stems also tend to be woodier. Two subspecies may be distinguished by their foliage characters. In subsp. *caucalifolium*, the leaves are green in colour, smooth or only slightly rough to touch and divided almost to the midrib with each segment also deeply divided. It is found in mountainous regions in southern and south-western Cape Province and was illustrated in 1794 by Jacquin from the gardens of the palace of Schönbrunn in Austria in 1794. The second subspecies, *convolvulifolium* (Schltr. ex Knuth) J.J.A. van der Walt, was first described as a new species by Knuth in 1912 and is distinguished by the rougher leaves which are more grey-green in colour and less deeply lobed. This is found in less mountainous areas. Neither is often seen in gardens but both have relatively large flowers making attractive low-growing, spreading subshrubs.

P. longicaule Jacquin, *Icones plantarum rariorum*, **3**: 10, t. 533 (1792)

MEANING: long-stemmed

SECTION: **Myrrhidium**

ILLUSTRATION: var. *longicaule* Sweet, **2**: 188 (a pink-flowered form); *Pelargoniums of southern Africa*, **1**: 28; var. *angustipetalum*, *Pelargoniums of southern Africa*, **3**: 84.

Low-growing spreading herbaceous plant stems woody at base. Leaves dark green, often with red tinge, deeply pinnately-divided to bipinnatisect, c. 5 × 4 cm (2 × 1 ^{3}/$_5$ in); petiole c. 5 cm (2 in); stipules broad ovate to 6 mm. Flowers white to pale pink, often tinged yellow in bud, on long branched flowering stems bearing small leafy bracts; peduncle usually 3–5 flowered, to 6 cm (2 ^{2}/$_5$ in); upper petals with conspicuous red feathering, broadly obovate, to 40 × 10 mm; lower petals 2 or 3, sometimes marked with red lines, lanceolate, to 20 × 4 mm; hypanthium to 5 cm (2 in); pedicel 1–2 mm; fertile stamens 7.

RECOGNITION: large flowers with 7 fertile stamens and finely-divided leaves.

The first reference to this plant is in 1767 when described as a form of *P. myrrhifolium* but there has been so much confusion about the nomenclature of members of this section that it is difficult to be sure if all the early botanists and gardeners were referring to the same plant. Even today many of the plants in cultivation as *P. myrrhifolium* var. *coriandrifolium* appear to belong to this species. It would however have been valued in the past for its flowers which are larger than most of the related species.

It may be seen growing in mountainous regions of south-western Cape Province and in cultivation forms an attractive plant for a hanging basket in cultivation where, in cooler climates, it will bloom through a long period of spring and summer. Plants with pinkish-yellow flowers and longer, but narrower upper petals to 4 cm (1 $^3/_5$ in) long may be referred to var. *angustipetalum* Boucher which is found in a limited area within the more northerly distribution of the species. This variety has only recently been recognized but would make a spectacular addition to any pelargonium collection.
(Plate 30)

P. multicaule Jacquin, *Icones plantarum rariorum*, 3: 10, t. 534 (1792)
MEANING: many stemmed
SECTION: **Myrrhidium**
ILLUSTRATION: subsp. *multicaule*, *Pelargoniums of southern Africa*, **2**: 96; subsp. *subherbaceum*, *Pelargoniums of southern Africa*, **3**: 95.

Low-spreading herbaceous plant, with woody older stems. Leaves almost triangular in shape with cordate base, deeply pinnately-divided with the segments themselves deeply and irregularly divided; petiole c. 3 cm (1 $^1/_5$ in); stipules ovate c. 5 mm. Flowers white to deep purplish-pink, in clusters of up to 5 flowers held well above the foliage on peduncles to 8 cm (3 $^1/_5$ in); upper petals obovate with long claws, 18 x 7 mm; lower petals narrower c. 10 x 3 mm; hypanthium 1–2 cm ($^2/_5$–$^4/_5$ in); pedicel 1–2 mm; fertile stamens 7, longer than the lower petals.

RECOGNITION: flowers with 4 petals; stamens exceeding lower petals.

Two subspecies have been designated in the recent revision. Plants with darker pink flowers, dark markings on upper petals, a hypanthium under 15 mm and leaf segments under 5 mm wide, often flushed red, are referred to subsp. *multicaule*. These plants are found over a wide area from the south-western to the eastern parts of Cape Province.

P. multicaule subsp. *subherbaceum* (Knuth) J.J.A. van der Walt, formally a species in its own right, differs in its white or pale pink tinged flowers, the upper petals with pale pink markings, the hypanthium of over 15 mm and broader leaf segments. It is found wild in the Transvaal and Natal. The Schönbrunn gardens of Emperor Joseph II near Vienna seem to have been the first to obtain this species towards the end of the eighteenth century but it is only grown nowadays in one or two specialist collections in Europe and North America. Although the flowers are relatively small, they are freely produced and like others of the section, this species makes a pretty plant in a hanging basket.

Herbarium specimens and descriptions of *P. whytei* Baker first described in 1897 and found in several countries of East Africa are very similar in appearance to *P. multicaule*. Two other species *P. gallense* Chiov. from Ethiopia and *P. goetzeanum* Engl. from Tanzania, are very similar and considered by some to be synonyms of *P. whytei*. Without seeing living, flowering plants, I prefer not to separate these species.

P. myrrhifolium (Linnaeus) L'Héritier, Aiton, *Hortus Kewensis*, **2**: 421 (1789)
MEANING: leaves similar to those of myrrh
SECTION: **Myrrhidium**
ILLUSTRATION: var. *myrrhifolium*, Sweet, **4**: 342; *Pelargoniums of southern Africa*, **1**: 27; var *coriandrifolium* Sweet, **1**: 34; *Pelargoniums of southern Africa*, **3**: 101.

Low-growing, short-lived sub-shrub, stems elongating when flowering. Leaves hairy or hairless, pinnatifid or bi-pinnatisect to about 7 x 4 cm (2 $^4/_5$ x 1 $^3/_5$ in) or more; petiole to 7 cm (2 $^4/_5$ in); stipules ovate, over 5 mm long. Flowers

P. alternans Wendland, J.C., *Hortus Herrenhausanus*, **1**: t. 10 (1798)

MEANING: alternate referring to leaf arrangement

SECTION: ***Otidia***

ILLUSTRATION: Sweet, **3**: 286 (somewhat stylized); *Pelargoniums of southern Africa*, **2**: 5.

Succulent forming a branching plant with stems becoming brown and woody with age. Leaves pinnate, segments deeply lobed, hairy, *c.* 40 × 15 mm; petiole *c.* 15 mm; stipules minute; Flowers usually white; inflorescence unbranched, hairy; upper petals narrow oblong, *c.* 12 × 2 mm, marked with red lines, reflexed; lower petals slightly smaller, sometimes marked with red, somewhat reflexed; hypanthium *c.* 7 mm; pedicel very short; fertile stamens 5, almost equal in length.

RECOGNITION: succulent with hairy, non-fleshy leaves; flowers with narrow, reflexed petals.

(actual size)

Old plants in the wild may be much more stunted and woody compared to those in cultivation which tend to retain their leaves and become more vigorous. It may be found in the drier and mountainous regions of south-western Cape Province and is said to have been introduced to Kew by F. Masson in 1791.

P. carnosum (Linnaeus) L'Héritier *Hortus Kewensis*, **2**: 421 (1789)

MEANING: succulent

SECTION: ***Otidia***

ILLUSTRATION: Sweet, **1**: 98; *Flowering plants of Africa*, **29**: 1145; *Pelargoniums of southern Africa*, **1**: 8.

Succulent plant to over 30 cm (12 in) with short, thickened, branched stems and a thickened rootstock. Leaves very variable, to 15 × 5 cm (6 × 2 in), usually less in cultivation, oblong, more or less succulent, often deeply pinnately-divided, grey-green to green; petiole 2–10 cm ($^4/_5$–4 in), swollen at base, usually persistent; stipules small, triangular; Flowers white sometimes with greenish tinge; inflorescence branched; upper petals *c.* 7 × 5 mm, broad ovate, slightly reflexed, marked with reddish-purple lines; lower petals slightly smaller, unmarked; hypanthium *c.* 6 mm; pedicel very short; fertile stamens 5, almost equal in length.

RECOGNITION: branching succulent with flattened pinnately-divided leaves.

This, the first known member of the section, was illustrated by Dillenius in 1732 and is still the most widely grown today. With its very wide natural distribution in southern and south-western Africa, it exhibits a great variation in morphological characteristics and has been collected on many expeditions with the result that there are many different synonyms.

Plants with very small petals have been named *P. parviflorum* (syn *P. brevipetalum* N.E. Brown) and some with rounded petals, distinguished by the unpublished name *P.* 'rotundipetalum', are perhaps best included within this variable species for the time being. The plant distinguished from others by its hairier foliage, less deeply divided leaves and conspicuous bracts is sometimes separated as a species, *P. polycephalum* (syn. *P. ferulaceum* var. *polycephalum*). Cultivated plants can be quite lush with large leaves looking quite unlike those collected in arid conditions.

P. ceratophyllum L'Héritier, *Hortus Kewensis*, **2**: 422 (1789)

MEANING: antler-like, referring to appearance of leaves

SECTION: ***Otidia***

ILLUSTRATION: *Botanical magazine*, **9**: 315; L'Héritier, *Geraniologia* 13; *Pelargoniums of southern Africa*, **2**: 29.

Succulent plant with thin branching stems. Leaves green to grey-green, succulent, pinnatifid to bipinnatifid, the ultimate segments grooved and almost cylindrical, c. 4 × 2 cm (1 ^{3}/$_5$ × ^{4}/$_5$ in); petiole to 4 cm (1 ^{3}/$_5$ in), persistent at base; stipules triangular; Flowers white to cream; inflorescence sometimes slightly branched; petals narrow elliptic to elliptic, c. 10 × 3 mm, the upper 2 crimped at base and feathered with purple-red, reflexed, slightly exceeding the conspicuous sepals; hypanthium to 1 cm (^{2}/$_5$ in); pedicel to over 1 cm (^{2}/$_5$ in); fertile stamens 5, more less equal in length.

RECOGNITION: small succulent with thin stems rarely exceeding 1 cm (^{2}/$_5$ in) diameter, ultimate leaf segments almost terete.

This is one of several species which grows wild in south-western Namibia and north-western Cape Province found in the region explored by A.P. Hove who collected these plants for the Royal Botanic Gardens of Kew in 1786. It is not difficult to cultivate and is found in some specialist collections.

P. crithmifolium Smith, J.E., *Icones pictae plantarum rariorum*, **1**: 13 (1793)

MEANING: leaves resembling samphire, *Crithmum maritimum*
SECTION: **Otidia**
ILLUSTRATION: *Botanical magazine*, **46**: 2029; *Flowering plants of Africa*, **45**: 1773; *Pelargoniums of southern Africa*, **1**: 11.

Branching succulent to about 50 cm (20 in) in cultivation, with rather thick knobbly greenish stems, sometimes covered with a peeling bark. Leaves fleshy, pinnately-divided into narrow, almost round segments channelled on the upper surface and toothed at the flattened apex; petiole about half length of lamina; stipules ovate. Flowers white on ends of dichotomously branched inflorescence; peduncles persistent becoming hard and thorn-like; upper petals, oblong, c. 12 × 5 mm, crimped with red markings at base; lower petals similar, unmarked but not crimped at base; hypanthium c. 2 mm; pedi-cel c. 2 cm (^{4}/$_5$ in); fertile stamens 5, of differing lengths.

RECOGNITION: branched succulent stems with branching thorn-like remains of persistent peduncles.

This species grows wild over a wide area of south-western Namibia and western Cape Province, usually in a rocky terrain. It was named by Sir James Smith who is thought to have received a plant or seeds collected by Captain Riou in 1790. It was however introduced into cultivation by F. Masson on his expedition, also in 1790, which judging by other species introduced at this time, must have been from the more southerly parts of its natural habitat. The persistent thorn-like remains of the inflorescence are said to provide some protection from grazing animals.

P. dasyphyllum Meyer, E., Original description in Knuth, *Das Pflanzenreich 4*, **129**: 373 (1912)

MEANING: means hairy leaf but hairs are not readily visible to naked eye
SECTION: **Otidia**
ILLUSTRATION: *Pelargoniums of southern Africa*, **2**: 48.

This species forms a branching succulent superficially similar to *P. crithmifolium* but with a less knobbly habit. The narrow, unmarked, strongly reflexed petals and stamens of equal length, give the flowers an asymmetric appearance. It has also been mistaken for *P. alternans*, but is less visibly hairy and has a less woody appearance. The flowers have a much more pronounced pedicel and the dichotomously branched inflorescence is more or less persistent. Found in the northwestern part of Cape Province and first collected in about 1830, it was included within *P. alternans* until distinguished formally by Knuth in 1912. *(Plate 32)*

P. klinghardtense Knuth, *Repertorium specierum novarum regni vegetabilis*, **18**: 292 (1922)

MEANING: name of mountain range where it was first found in Namibia

SECTION: ***Otidia***

ILLUSTRATION: *Pelargoniums of southern Africa*, **2**: 84.

Succulent branching from base with swollen blue- or yellowish-green stems. Leaves succulent, blue-green, obovate, apex lobed or toothed, sometimes wavy; sometimes with two lobes at base, narrowing into the very short petiole which is swollen at its base; stipules minute. Flowers white, inflorescence regularly branched, with yellowish-green peduncles, often persistent; all petals more or less similar, unmarked, spathulate, to 12 x 4 mm, slightly exceeding conspicuous yellow green sepals; hypanthium *c.* 5 mm; pedicel *c.* 15 mm; fertile stamens 5, more or equal.

RECOGNITION: simple leaves and short swollen stems.

First discovered on an expedition made by Gordon and Paterson northwards from the Cape of Good Hope along the western coast of South Africa in 1778 and 1779, this plant was not formally recognized until 1922 despite illustrations made by Gordon nearly 150 years earlier. It grows in southern Namibia and is occasionally grown as a curiosity in gardens.

P. laxum (Sweet) Don, G., *A general history*, **1**: 730 (1831)

MEANING: spreading open appearance, referring to inflorescence

SECTION: ***Otidia***

ILLUSTRATION: *Pelargoniums of southern Africa*, **3**: 83.

Branching succulent, stems yellowish-green with peeling bark in older plants. Leaves deeply pinnatifid, segments deeply, irregularly toothed, ovate, to over 10 x 5 cm (4 x 2 in), thickened but not succulent; petiole *c.* 6 cm (2 $^2/_5$ in); stipules triangular. Flowers white, very irregular in appearance; inflorescence repeatedly branched; upper petals crimped at base with reddish markings, strongly reflexed from base, oblong, to 15 x 4 mm; lower petals unmarked with claw; hypanthium *c.* 5 mm; pedicel *c.* 15 mm; fertile stamens 5, unequal in length.

RECOGNITION: branching succulent with large branching inflorescence similar to *P. paniculatum* but leaf bases not persistent on stems and with flattened laminas.

The only species of the section to be found outside the western side of South Africa and Namibia, *P. laxum* grows in the summer rainfall zones of central and eastern Cape Province. It is said to have been cultivated by Colvill from seed sent back from the Cape in 1821.

P. paniculatum Jacquin, *Plantarum rariorum horti caesarei Schoenbrunnensis*, **2**: 6, t. 137 (1797)

MEANING: repeatedly branched inflorescence

SECTION: ***Otidia***

ILLUSTRATION: *Pelargoniums of southern Africa*, **2**: 108.

Plant succulent, stem greenish, columnar, rarely branched but covered with the persistent bases of the petioles creating a spiny effect. Leaves greyish-green, to over 10 x 9 cm (4 x 3 ³/₅ in), 2–3 pinnate, ultimate segments linear, almost round; petiole to over 10 cm (4 in), swollen at base; stipules ovate. Flowers cream to white, very irregular in appearance; inflorescence repeatedly branched; peduncles and sepals often tinged red; upper petals *c.* 10 x 3 mm, oblong, marked with purplish-red at the crimped base, reflexed; lower petals similar to upper but unmarked and not crimped; hypanthium *c.* 1 mm; pedicel *c.* 12 mm; fertile stamens 5, unequal in length.

RECOGNITION: unbranched succulent stems covered with persistent leaf bases, large inflorescence repeatedly branched.

This species, found wild in a small area of the very hot dry region of south-western Namibia, was collected at the end of the eighteenth century. It is not difficult to grow and in flower is quite spectacular reaching to well over 50 cm (20 in).

Pelargonium de Candolle

This is a large section, named for the genus itself. Some species will appear very familiar. It includes the parents of most of the scented-leaved cultivars, the angel pelargoniums and the Uniques, as well as the species, *P. cucullatum*, from which all the regal pelargoniums have been developed. A recent monograph of the section recognizes 24 species, most of which are in cultivation.[21]

These species may usually be found in areas where there is at least some water, in the subsoil or near streams or rivers, even though the latter are often dry for part of the year. The majority are to be found in the winter rainfall regions of south-western or southern Cape Province, many near to the coast or by well-trodden routes which explains why so many were known to the early botanists and collectors.

When allowed a free-root run and left unpruned, most will become large and eventually, quite woody. The foliage is often aromatic and sometimes viscid which must contribute in some way to their survival in their natural habitat. The flowers are in rather few-flowered inflorescences and white, or in shades of pink or purple, but none are a true red or yellow. The upper petals differ from the lower in shape, are larger in size and are marked with darker spots or lines. There are 7 fertile stamens and the hypanthium may be longer or shorter than the pedicel. The basic chromosome number is $x = 11$, and most have 22 or 44 rather small chromosomes. *P. radens*, *P. vitifolium* and *P. graveolens* have 88 chromosomes and *P. capitatum* has 66. They appear closely related to the section *Glaucophyllum* but do not grow in such arid situations.

All are easy to grow as long as they are not overwatered and are given sufficient light. If planted outside in frost-free climates most will make large bushy plants but even in cooler areas may be planted out, making a useful summer addition to a scented garden or container. Cuttings may be taken towards the end of the summer. Large plants can be cut back quite hard without damage and potted up to be kept in frost-free conditions for overwintering. Although not as suitable as zonal cultivars for a formal bedding scheme because of their less tidy habit, they are very colourful and long flowering in informal situations.

If seed is collected, it is important to remember that some do hybridize and the resulting plants may be unexpected. Hybrids are found in the wild and in cultivation, not only between species within this section but also with members of some other sections such as *Glaucophyllum*. The cultivars grown for rose-scented geranium oils have been developed from species such as *P. capitatum* and *P. radens*, and several other species have scented foliage which have been considered as possible sources of aromatic oils. The scented foliage may be used in pot-pourri and some are used for culinary flavouring. This section therefore contains some very important plants for both the horticultural industry and as a crop for the commercial production of geranium oil.

1	Leaves viscid	2
	Leaves not viscid	5
2	Leaves not aromatic	**pseudoglutinosum**
	Leaves aromatic	3
3	Leaves rough	4
	Leaves not rough	**glutinosum**
4	Leaf segments usually under 3 mm wide, grooved along midrib	**denticulatum**
	Leaf segments usually over 3 mm wide, not grooved along midrib	**quercifolium**
5	Leaves pinnatisect, pinnae also deeply divided	6
	Leaves not so deeply or repeatedly divided	7
6	Inflorescence lax; leaf segments about 3 mm wide	**radens**
	Inflorescence compact; leaf segments over 4 mm wide	**'Graveolens'**
7	Pedicel under 3 mm long	8
	Pedicel over 3 mm long	10
8	Upper petals over 2 cm ($^4/_5$ in) long	**panduriforme**
	Upper petals under 2 cm ($^4/_5$ in) long	9
9	Leaves soft, rose-scented	**capitatum**
	Leaves rough, somewhat lemon-scented	**vitifolium**
10	Leaves peppermint-scented	**tomentosum**
	Leaves not peppermint-scented	11
11	Leaves lemon-scented	12
	Leaves not lemon-scented	14
12	Leaves under 2 cm ($^4/_5$ in) across	**crispum**
	Leaves over 2 cm ($^4/_5$ in) across	13
13	Inflorescence frequently branched	**citronellum**
	Inflorescence rarely branched	**scabrum**
14	Lobes of leaves acute	15
	Lobes of leaves obtuse or leaves unlobed	16
15	Pedicel longer than hypanthium; young stems green	**hispidum**
	Hypanthium longer than pedicel; young stems red	**scabroide**
16	Upper petals over 2 cm ($^4/_5$ in) long	17
	Upper petals not exceeding 2 cm ($^4/_5$ in), usually less	19
17	All petals more or less similar in shape and size	18
	Upper and lower petals very dissimilar in shape and size	**cordifolium**
18	Leaves usually over 4 cm (1 $^3/_5$ in) across, often cupped and softly-hairy	**cucullatum**
	Leaves usually under 3 cm (1 $^1/_5$ in) across, never cupped not softly-hairy	**betulinum**
19	Flowers white	**ribifolium**
	Flowers pink	19
20	Leaves arranged in two opposite rows	**hermanniifolium**
	Leaves not arranged in two opposite rows	21
21	Plant robust to over 1 m (40 in) tall, leaves with unpleasant scent	**papilionaceum**
	Plant not exceeding 70 cm (28 in), without unpleasant scent	22
22	Leaves crisped with rose-like scent	**englerianum**
	Leaves not crisped and not rose-scented	23
23	Young stems green	**greytonense**
	Young stems brown to reddish	**sublignosum**

P. betulinum (Linnaeus) L'Héritier, Aiton, *Hortus Kewensis*, **2:** 429 (1789)

MEANING: birch-like, referring to the leaves resembling *Betula* in shape

SECTION: ***Pelargonium***

ILLUSTRATION: *Botanical magazine*, **5:** 148; *Pelargoniums of southern Africa*, **1:** 5.

Upright or decumbent branching plant, woody at base. Leaves rather stiff, sometimes roughly hairy, ovate to rounded, toothed, the teeth sometimes red-tipped, c. 2 × 1.5 cm (⁴/₅ × ³/₅ in); petiole c. 1 cm (²/₅ in); stipules triangular. Flowers pink or purplish-pink, sometimes white, 3 cm (1 ¹/₅ in) or more across; peduncle 2–8 cm (⁴/₅–3 ¹/₅ in) usually with 3–4 flowers; upper petals heavily veined with purplish-red, broadly obovate, to 25 × 15 mm; lower petals unmarked or with faint markings, narrower than upper; hypanthium c. 5 mm pedicel to 15 mm, exceeding hypanthium; fertile stamens 7.

RECOGNITION: small woody plant with large flowers in few-flowered heads.

This species which is found near the south-western and southern coast line of South Africa, was known in European gardens in the mid-eighteenth century and cultivated by Phillip Miller as early as 1759. The large flowers would have made it an obvious choice of parent for hybridizing although in the early days before the mechanics of pollination were properly understood, crosses were accidental. It is almost certainly one of the ancestors of some of the regal and Unique pelargoniums and this is quite plausible as natural hybrids with P. cucullatum are found in the wild, although nowadays it is surprisingly rare in collections. It has been used for the treatment of chest complaints in its native land. (Plate 33)

P. capitatum (Linnaeus) L'Héritier, Aiton, *Hortus Kewensis*, 2: 425 (1789)

MEANING: head, referring to compact inflorescence

SECTION: **Pelargonium**

ILLUSTRATION: *Pelargoniums of southern Africa*, **1**: 7.

Decumbent, somewhat spreading, or weakly erect, softly-hairy plant with stems becoming more or less woody at base, to about 1 m (40 in) in height. Leaves rose-scented, densely hairy and somewhat crinkled with rounded lobes to 2–8 cm (⁴/₅–3 ¹/₅ in) across; petiole to 4 cm (1 ³/₅ in); stipules cordate. Flowers to 15 mm across, usually mauve-pink in very dense compact heads of 10–20 flowers; peduncle erect to 10 cm (4 in); upper 2 petals narrow obovate, veined darker pink, c. 18 × 5 mm; lower petals slightly smaller, less marked; hypanthium to 3 mm; pedicel to 1 mm or less; fertile stamens 7. Fruit has a characteristic spiny appearance due to the very short pedicel and hypanthium and swollen mericarps.

RECOGNITION: hairy decumbent, rose-scented plant distinguished from *P. vitifolium* by habit and scent.

P. capitatum was one of the earlier species imported to Europe. Records indicate that it was brought into England from Holland in 1690 by Hans Willem Bentinck, the head steward to William of Orange who later became the 1st Earl of Portland. It is not surprising that this was such an early introduction as it is found in many areas along the coasts of South Africa easily accessible to explorers or sailors. As well as the attractive flowers, it has strongly rose-scented foliage. However, some plants may be found that are more erect in their habit with less hairy leaves and more star-shaped flowers. It appears, although a definitive study has not been carried out, that these types have less scented leaves and there are some in cultivation, at least in the United Kingdom, which have caused confusion in the nomenclature of the species. It often grows in sandy dunes in areas where the rain falls either in winter, in summer or in both seasons.

The only other species that might be confused in flower with *P. capitatum* is *P. vitifolium*, also one of the early introductions. This is a much more vigorous, erect plant with rougher, less crinkled leaves and a harsh almost unpleasant scent.

The true species is quite rare in gardens but is represented by the cultivar 'Attar of Roses' with a more upright habit and rougher but strongly aromatic leaves and pinker flowers. The anthers bear little pollen and fertile seed is rarely set so it is presumably a hybrid of the species.

'Otto of Roses' is a form of *P. capitatum* mentioned by Andrews in his book *Geraniums* in 1805. He describes a similar plant that is 'easier to keep' and it would appear that this might be the same or very similar to the plant known today. There are other cultivars found in gardens derived from *P. capitatum* including 'Atomic Snowflake', with leaves streaked and blotched with white.

P. citronellum J.J.A. van der Walt, *South African journal of botany*, **49:** 79 (1983)

MEANING: citron referring to scent of foliage

SECTION: **Pelargonium**

ILLUSTRATION: *Pelargoniums of southern Africa*, **3:** 29.

Branching, roughly hairy, lemon-scented shrub to 2 m (6 $^1/_2$ ft). Leaves palmatisect or palmatifid, lobes irregularly toothed, base cordate and apices acute, *c.* 5 x 5–6 cm (2 x 2–2 $^2/_5$ in); petiole 3–4 cm (1 $^1/_5$–1 $^3/_5$ in); stipules triangular. Flowers pale pinkish-purple in open branching inflorescence, each peduncle of about 5 flowers; upper petals obovate, reflexed, marked with dark purple lines and blotches, to 20 x 7–8 mm; lower petals unmarked, smaller, oblong, unmarked; hypanthium *c.* 5 mm; pedicel *c.* 10 mm; fertile stamens 7.

RECOGNITION: robust plant with lemon-scented leaves, distinguished from *P. scabrum* by larger flowers and pedicels exceeding hypanthium.

Until recently, this plant was not recognized as a distinct species because of its similarity to *P. scabrum*. It normally grows near streams in a small area of southern Cape Province. In cultivation it forms a very robust plant and sets seed freely but cuttings may sometimes be slow to root. It is advisable to try to encourage branching by pinching out the shoot tips, as in cultivation there is a tendency for the main stem to grow upwards with few branches. Several cultivars such as 'Mabel Grey' show a very close resemblance to the species in morphology as well as in their habit and behaviour in cultivation, and may be selections or hybrids of it or of *P. scabrum*. (Plate 34)

P. cordifolium (Cavanilles) Curtis, *Botanical magazine*, **5:** 165 (1791)

MEANING: heart-shaped leaves

SECTION: **Pelargonium**

ILLUSTRATION: *Botanical magazine*, **5:** 165; *Pelargoniums of southern Africa*, **1:** 9.

Branching hairy shrub to 1.5 m (5 ft). Leaves rather dark grey-green, undersurface hairy, heart-shaped, sometimes with shallow lobes, toothed, to 6 x 5 cm (2 $^2/_5$ x 2 in); petiole 2–3 cm ($^4/_5$–1 $^1/_5$ in); stipules broad-triangular. Flowers purplish-pink to pink, very irregular in appearance in heads of up to 8 flowers in a branching inflorescence; upper petals veined dark purple, obovate, reflexed, 25–30 x 10 mm; lower petals paler, unmarked, linear; hypanthium 5–6 mm; pedicel slightly shorter than hypanthium; fertile stamens 7.

RECOGNITION: heart-shaped, sometimes lobed leaves; flowers with large showy upper petals but very narrow lower ones.

This is one of the many species introduced by F. Masson in 1774 and is found along a wide area of the southern part of South Africa. The under-surface of the leaves of plants collected in the wild is variable, those of some plants having very silvery-white felted appearance while in others the surface is only slightly hairy. Those with very hairy leaves have in the past been given the varietal name var. *lanatum*, while others with a distinctive red margin to the leaves have been named var. *rubrocinctum*. In cultivation, it is probably best to treat distinctive vegetatively maintained plants as cultivars. Unless the plant is growing freely in the ground, it has the tendency to grow upwards with little branching so it is advisable to pinch out the growing shoot to encourage a more bushy habit. *(Plate 35)*

P. crispum (Bergius) L'Héritier, Aiton, *Hortus Kewensis*, **2**: 430 (1789)

MEANING: crisped, referring to margins of leaves
SECTION: **Pelargonium**
ILLUSTRATION: Sweet, **4**: 383; *Pelargoniums of southern Africa*, **3**: 36.

Erect plant to 70 cm (28 in) high, with stems becoming woody at base. Leaves rough, strongly lemon-scented, margins crisped, to about 7 mm across; petiole c. 5 mm; stipules heart-shaped, conspicuous to 4 mm long. Flowers usually pink, large; peduncle 1–2 flowered, to 10 mm; upper petals spathulate, with darker markings, apex notched, c. 20 × 10 mm; lower petals narrow oblong, unmarked; hypanthium to 8 mm; pedicel c. 5 mm; fertile stamens 7.

RECOGNITION: erect lemon-scented shrub with large flowers but very small leaves. Morphologically similar to *P. hermanniifolium* which is rarely aromatic and *P. englerianum* which has larger leaves but much smaller flowers.

(actual size)

Garden plants show a wide range of leaf size. Vigorous cultivars with distinctly larger leaves are grown such as 'Major' and the slightly smaller leaved, 'Minor'. The true species with very tiny crisped leaves is a less vigorous plant but always produces an abundance of fertile seed. It has been known in America as the finger bowl pelargonium. Those with leaves edged with creamy-white are named 'Variegatum'. The species grows wild in south-western Cape Province and was introduced to Kew gardens in 1774 by F. Masson but must have been known even earlier as Bergius, the pupil of Linnaeus, had described it seven years before.

P. cucullatum (Linnaeus) L'Héritier, Aiton, *Hortus Kewensis*, **2**: 426 (1789)

MEANING: cup-like, referring to leaves
SECTION: **Pelargonium**
ILLUSTRATION: *Pelargoniums of southern Africa*, **1**: 12; **2**: 42; **3**: 41.

Erect branched, hairy, shrub to over 2 m (6 $^1/_2$ ft) in height. Leaves rounded to triangular in shape, toothed and sometimes shallow-lobed, usually hood- or cup-shaped with long hairs, often with a thin reddish margin, and sometimes aromatic, usually c. 8 × 7 cm (3 $^1/_5$ × 2 $^4/_5$ in); petiole to c. 2 cm ($^4/_5$ in); stipules ovate, to 10 mm. Flowers to over 4 cm (1 $^3/_5$ in) across, bright purplish-pink in large branched inflorescence, each head with about 5 flowers; peduncle to c. 5 cm (2 in); upper 2 petals broad obovate, veined deeper pink, to 25 × 15 mm; lower petals slightly smaller, narrower, unmarked; hypanthium 5–12 mm; pedicel to 10 mm; fertile stamens 7.

RECOGNITION: large shrub with more or less cup-shaped leaves; flowers large, purple to pinkish-purple.

P. cucullatum grows wild in Cape Province but there are three subspecies, for many years treated as separate species, found in distinct geographical areas which rarely overlap.[22] The characteristics of the three subspecies in the wild are quite easily defined and Dillenius in 1732 recognized more than one different plant. Linnaeus, however, grouped them all together. *P. cucullatum* subsp. *cucullatum* grows mainly to the east of False Bay in the coastal regions from very near the beach to the foothills of the mountains, on well-drained and sandy soils. This was introduced into cultivation in the early part of the eighteenth century, grown at Chelsea Physic Garden in 1724 and illustrated as early in 1732.

It is also probable that this or subsp. *strigifolium* is one of the original parents of the regal, fancy or Martha Washington pelargoniums, known botanically as **P × domesticum** Bailey. The leaves are angular in shape, hence the early specific name *P. angulosum*, cupped at the base and covered with long soft hairs. The flowers are usually dark pinkish-purple and large to about 4 cm (1 $^3/_5$ in) across.

P. cucullatum subsp. *strigifolium*, Volschenk, a less showy plant also has angular leaves but these are less cupped and much rougher to the touch. The flowers are somewhat smaller and tend to be paler in colour but it crosses more readily with other species than do the other two subspecies. It grows further inland and at higher altitudes. This was known in European gardens as *P. acerifolium* and cultivated by Mr Archibald Thompson in 1784.

P. cucullatum subsp. *tabulare* Volschenk is found mainly on the Cape Peninsula in similar conditions to those favoured by subsp. *cucullatum*. In this subspecies the leaves are rounded in shape without angular incisions, softly-hairy and distinctly cup-shaped. They are also strongly scented, a fact that has barely been mentioned in the literature except by the earliest botanists. The flowers are large and deep pinkish-purple. It would seem possible judging from the scent, softness and shape of the foliage as well as the large, deep purple flowers that it was from this plant that some hybrids such as 'Purple Unique' have been developed. It is considered that this is probably the one first introduced into England in 1690 by Mr Bentinck and also described earlier by Hermann from Leiden in 1687 as *Geranium africanum arborescens, ibisci folio rotundo, carlinae odore*.

In cultivation it is often difficult to decide which subspecies to assign a plant, probably because of the hybridization of the subspecies. Natural hybrids have been found between *P. cucullatum* and *P. betulinum* which may prove to be useful garden plants as well as possible clues to the parentage of some cultivars of uncertain origin. Plants with double or white flowers and others with variegated leaves have been in cultivation in the past and records may be found in the older literature. One, 'Flore Plenum', with semi-double flowers is still grown today.
(Plates 36 & 37)

P. denticulatum Jacquin, *Plantarum rariorum horti caesarei Schoenbrunnensis*, **2**: 5, t. 135 (1797)

MEANING: toothed referring to leaves
SECTION: **Pelargonium**
ILLUSTRATION: Sweet **2**: 109; *Pelargoniums of southern Africa*, **2**: 53.

Erect branched shrub to over 1.5 m (5 ft), strongly balsam-scented, sticky. Leaves very sticky, dark green sometimes with darker markings especially along veins, triangular in shape, base cordate, bipinnatisect, ultimate segments irregularly toothed, c. 6–8 × 7–9 cm (2 $^2/_5$–3 $^1/_5$ × 2 $^4/_5$–3 $^3/_5$ in), petiole c. 5 cm (2 in); stipules triangular. Flowers purplish-pink, to 2 cm ($^4/_5$ in) across; peduncle to 5 cm (2 in) bearing about 6 flowers; upper petals narrow spathulate, veined darker pink, apex usually notched, c. 18 × 6 mm; lower petals slightly smaller, unmarked; hypanthium c. 8–9 mm; pedicel to 2 mm; fertile stamens 7.

RECOGNITION: shrub with sticky, finely divided leaves. Distinguished from other viscid species by the foliage which is similar in shape to that of *P. radens*, but this latter species has rougher, non sticky leaves.

It may be found in south-western and southern Cape Province growing in the moister areas

P. cucullatum subsp. *tabulare*

'Filicifolium'

especially beside streams and in deep valleys. It was introduced into cultivation at Kew in 1789. Also known in cultivation for a very long time is a form of this species with leaves so finely divided that it has been given the name 'Filicifolium'. An additional feature of this cultivar is that the upper two petals are deeply notched, often divided to almost half their length. Leaves of plants within a single population in the wild may show different degrees of division and some resemble 'Filicifolium', but the plant in cultivation has been maintained as a distinct form for many years and justifies the application of a cultivar name.

P. englerianum Knuth, *Das Pflanzenreich* 4, **129**: 470 (1912)

MEANING: named for A. Engler, an important German taxonomist
SECTION: **Pelargonium**
ILLUSTRATION: *Pelargoniums of southern Africa*, **3**: 53.

Erect or spreading, aromatic subshrub with purple or brown stems covered with long hairs. Leaves rough, crisped, or wavy-edged, reniform, sometimes shallow-lobed, margin with large teeth, *c.* 2 x 3 cm ($^4/_5$ x 1 $^1/_5$ in); petiole with long hairs, *c.* 15 mm; stipules cordate. Flowers pinkish-purple; peduncle to 4 cm (1 $^3/_5$ in) bearing up to 5 flowers; upper petals spathulate with dark purplish-red marks and lines, *c.* 15 x 7 mm; lower petals narrower, clawed; hypanthium to 10 mm conspicuously swollen at base; pedicel exceeding hypanthium; fertile stamens 7.

RECOGNITION: subshrub with crisped leaves scented of a mixture of rose and camphor and small pinkish-purple flowers.

This relatively unknown species grows in inland mountainous areas of south-western Cape Province. It is not difficult to cultivate and the aromatic foliage and freely produced though small flowers make it an attractive addition to a collection of species or scented-leaved pelargoniums.

P. glutinosum (Jacquin) L'Héritier, Aiton, *Hortus Kewensis*, **2**: 426 (1789)

MEANING: sticky, referring to leaves
SECTION: **Pelargonium**
ILLUSTRATION: *Botanical magazine*, **4**: 143; *Pelargoniums of southern Africa*, **3**: 59.

Erect, branching, viscid, shrub to well over 1 m (40 in) in height, strongly scented of balsam. Leaves variable in shape, rather dark green, broadly triangular in shape, palmately 3-lobed, but sometimes divided almost to midrib, lobes irregularly and coarsely toothed with acute apices, leaf base cordate, c. 5 × 5.5 cm (2 × 2 1/5 in) but often considerably larger if grown in open ground; petiole c. 3 cm (1 1/5 in); stipules triangular or narrow triangular. Flowers pale to dark pink, c. 1.5 cm (3/5 in) across; inflorescence 1–8 flowered, peduncle to 8 cm (3 1/5 in); upper 2 petals spathulate with darker markings, usually c. 13 × 6 mm but variable in size; lower petals unmarked; hypanthium c. 10 mm; pedicel very short; fertile stamens 7.

RECOGNITION: Large shrub-like plant with very sticky leaves distinguished from *P. denticulatum* by the leaves which are never bipinnatisect.

P. glutinosum has a far wider distribution in the wild than some of the related species with viscid leaves, but tends to inhabit similar rather moist situations. It shows a range of variation in both its leaf and flower characteristics but the leaves are never so finely or regularly divided as those of *P. denticulatum*. Plants from the western part of Cape Province have the most divided leaves and the plant with deeply divided leaves, illustrated by Sweet in 1822 (**2**: 118) as 'Viscosissimum' is probably a form of the species. There are several other plants to be found in gardens such as 'Pheasant's Foot', and 'Asperum' which may be forms of *P. glutinosum* or hybrids with *P. denticulatum*. The flowers have a more regular appearance than many related species, the upper petals being less reflexed and the buds rather bulbous in shape.

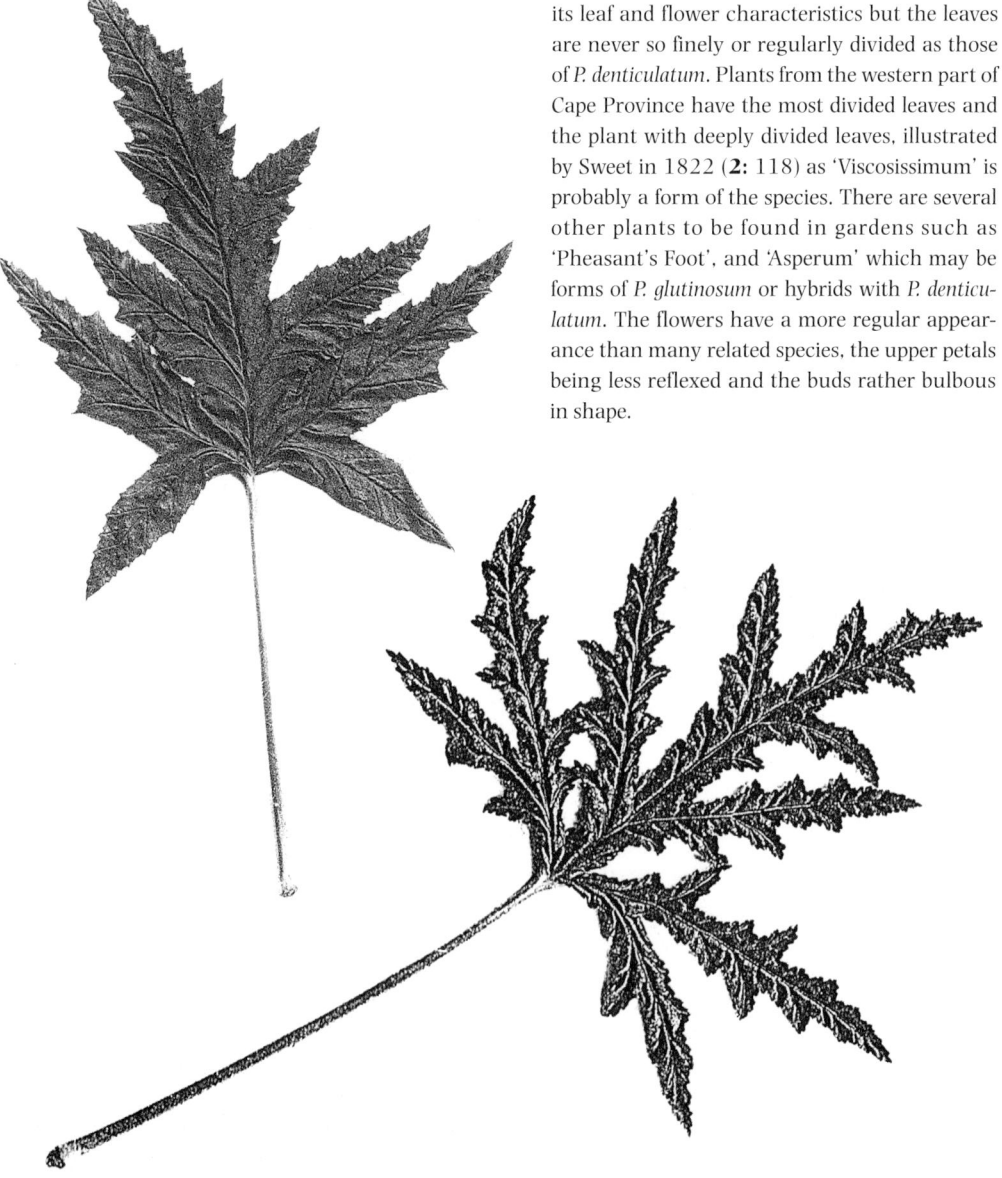

P. 'Graveolens'

MEANING: strongly scented
SECTION: **Pelargonium**
ILLUSTRATION: L'Héritier, *Geraniologia* t. 17 (1792).

Vigorous, erect, branching, mint to rose-scented, shrub to over 1.5 m (5 ft) in height. Leaves somewhat rough, green, with segment margins somewhat curved under, *c.* 4 × 6 cm (1 $^{3}/_{5}$ × 2 $^{2}/_{5}$ in); petiole *c.* 3 cm (1 1/5 in); stipules broad-triangular, 5–8 mm. Flowers pink to pale purplish-pink, to 15 mm across; peduncle to 4 cm (1 $^{3}/_{5}$ in) bearing 5–10 flowers; upper petals narrow obovate, with rounded or notched apex, veined dark purplish-pink, *c.* 10–15 × 5 mm; lower petals smaller, unmarked; hypanthium *c.* 5 mm; pedicel to 3 mm; stamens with stunted anthers rarely bearing pollen.

RECOGNITION: robust, rose or minty-rose-scented plant with pink flowers in tight few flowered heads.

P. 'Graveolens'

The naming of the old and exceedingly well-known rose-scented pelargonium grown on the window sills of so many houses in this country, Europe and North America, is possibly the most difficult to resolve. In 1792, L'Héritier illustrated a plant and named it *P. graveolens*. The plant depicted in this plate is more or less identical morphologically with the plant most commonly grown today under this name in gardens and it was almost certainly a hybrid. It is obviously not possible to compare scents or even the colour accurately from a non-coloured engraving but the leaf shape, texture, habit, stipules, inflorescence structure, flower size, shape and markings and even the lack of large apparently fertile anthers are very clear.

The present day plant is clearly a hybrid which does not readily set seed. Occasionally fertile seed is set producing plants with minor differences in character such as leaf shape and scent. Some of these have resulted in named ornamental cultivars such as 'Little Gem' with a more compact habit and slightly larger rose-pink flowers. 'Lady Plymouth' with leaves margined cream must have been a sport. A third cultivar considered incorrectly for many years a species, *P. radula* (see leaf silhouette under *P. radens* on page xxx), with harsher, more divided and more rose-scented leaves, is very similar. It is best treated as a cultivar ,'Radula', and not confused with the species now known as *P. radens*. All these, and several others are related to the original 'P. graveolens' L'Hérit., and aredistinguished by variations in habit, leaf shape or scent, but none of these have truly fertile anthers.

Some plants may produce limited seed when cross-pollinated. The inclusion of all these related cultivars, as well as those such as 'Rosé', grown commercially for the extraction of geranium oil, in a cultivar group, Graveolens, would help to distinguish these from the plant illustrated in *Pelargoniums of South Africa*, **3**: p.63 (1988) as *P. graveolens*. This is a species only seen in specialist collections and is found wild in the northern part of South Africa as well as in Zimbabwe and Mozambique. It is unlikely to have been known in the late eighteenth century. This has white flowers and deeply divided, softly-hairy, somewhat peppermint-rose-scented leaves. The flowers are larger, fewer with longer pedicels and hypanthia giving the inflorescence a more open appearance. It also bears seven fertile anthers, and produces abundant seed which germinates readily to produce similar plants. This is a quite distinct species and it should be given a new specific name to avoid confusion. There is no need for the cultivar name 'White Form', which has recently been incorrectly applied, at least in the United Kingdom. The plants grown for over 200 hundred years as *P. graveolens*, under the rules of the International

Code of Nomenclature for Cultivated Plants, can quite legitimately retain the old specific epithet as a cultivar name.

Several cultivars related to 'Graveolens' have been grown in very large quantities for the production of rose-scented geranium oil in many regions of the world. Different regions have developed their own, each resulting in oils with slightly differing characteristics. 'Rosé', cultivated on Reunion Island, has been subjected to intensive studies including analyses of the essential oils and chromosome counts, by researchers in South Africa and Reunion Island. The results indicate that the parents of 'Rosé' are *P. capitatum* (2n = 66) and *P. radens* (2n = 88). It is interesting to add that chromosome counts made on 'Graveolens' are the same as for 'Rosé' (2n = 77). It is also significant that in South Africa plants were found in the wild, in the vicinity of both *P. capitatum* and *P. radens*, with similar morphological characteristics to some of the known cultivars of Graveolens Group.

Deliberate crosses of the two presumed parents also produce a range of similar seedlings with slightly differing scents and foliage. There is a great deal of confusion in the nomenclature, relationships and parentages between all these plants, and as many old established cultivars and new deliberate hybrids as possible are being assembled at the N.C.C.P.G. National Collection of Pelargoniums for further study.
(plate 38 and line drawing 11)

P. greytonense J.J.A. van der Walt, *South African journal of botany*, **50:** 256 (1984)

MEANING: from village where it is found
SECTION: **Pelargonium**
ILLUSTRATION: *Pelargoniums of southern Africa*, **3:** 66.

An upright branching, sometimes aromatic shrub eventually reaching about 1 m (40 in) in height. Leaves shallow 3-lobed, apex rounded, margins coarsely toothed, base cordate to truncate; petiole *c.* 5 cm (2 in); stipules broad-triangular. Flowers white or pale pink; inflorescence to 9-flowered with leafy peduncle to 7 cm (2 $^4/_5$ in); upper petals narrow obovate, with dark red marks and lines, *c.* 20 × 8 mm; lower petals narrower; hypanthium *c.* 5 mm; pedicel to 20 mm; fertile stamens 7.

RECOGNITION: upright plant with attractively marked pink or white flowers.

P. graveolons

This newly named species is found in a small area of south-western Cape Province and it has been suggested that it may have arisen originally as a hybrid between *P. papilionaceum* and *P. hermanniifolium*. It flowers freely in cultivation and may become a useful addition to collections.

Fig. 11   P. 'Graveolens' from L'Héritier, *Geraniologia* (1792)

P. hermanniifolium (Bergius) Jacquin, *Icones plantarum rariorum*, **3**: 11, t. 545 (1792)

MEANING: leaves like *Hermannia*, an African genus of plants

SECTION: ***Pelargonium***

ILLUSTRATION: *Pelargoniums of southern Africa*, **2**: 71.

With its stiff, erect habit, large pink flowers and rather small leaves, this species is similar to and has been confused with *P. crispum* ever since its first recognition in 1767. It grows in approximately the same geographical region but in areas of slightly higher rainfall. It may be distinguished by the leaves which are strictly arranged in two opposite rows along the stems. The leaves are obovate with a cuneate, not cordate base and they are never lemon-scented. It is not easy to root but does make an attractive small plant. *(Plate 39)*

P. hispidum (Linnaeus f.) Willdenow, *Species plantarum*, **3**: 677 (1800)

MEANING: rough, referring to hairs on leaves

Section: ***Pelargonium***

ILLUSTRATION: *Pelargoniums of southern Africa*, **2**: 72.

Erect, branched, aromatic shrub to over 2 m (6 $^{1}/_{2}$ ft) in height covered with rough hairs and bristles. Leaves palmatifid, 5–7 lobed, irregularly and coarsely toothed, apices acute, base cordate, conspicuously veined, about 10 x 15 cm (4–6 in) but may become larger in cultivation; petiole 12–20 cm (4 $^{4}/_{5}$ x 8 in); stipules triangular. Flowers to 1.5 cm (6 in) across, pale to dark pink in branched inflorescence held above the foliage, each cluster 6–12 flowered on peduncles to *c.* 4 cm (1 $^{3}/_{5}$ in); upper 2 petals reflexed, obovate, with dark red marks, *c.* 12 x 7 mm; lower petals much smaller, to 8 mm long, narrow; hypanthium *c.* 4 mm; pedicel to 7 mm; fertile stamens 7.

RECOGNITION: Tall growing shrub with rough maple-shaped leaves with acute apices; flowers somewhat similar to *P. papilionaceum* with very showy upper petals but insignificant lower ones.

With a free root run this species will become a very large shrub with attractive foliage, but it is not often grown, perhaps because in a pot it is less significant and does not have such a pleasant scent or as showy flowers as many other species of pelargonium. It was probably collected from the mountainous regions of south-western Cape Province on several occasions but the only record of it is from F. Masson who brought back specimens for Kew in about 1790 although it had been described and illustrated some years before this date.

P. panduriforme Ecklon & Zeyher, *Enumeratio plantarum*, 1: 82 (1835)

MEANING: refers to leaf shape

SECTION: ***Pelargonium***

ILLUSTRATION: *Pelargoniums of southern Africa*, **3**: 108.

Erect, branched, shrub to over 2 m (6 $^{1}/_{2}$ ft) in height, strongly balsam-scented, sometimes slightly viscid and with long hairs. Leaves soft with long hairs, pinnately lobed, the lobes with very shallow broad scalloped teeth with rounded apices, base cordate, *c.* 3.5 x 2.5 cm (1 $^{2}/_{5}$ x 1 in) but when grown in the open ground may be 2 or 3 times larger; petiole to 2 cm ($^{4}/_{5}$ in) or more, stipules triangular. Flowers large, pink, in inflorescence of 2–20 flowers; peduncle to 8 cm (3 $^{1}/_{5}$ in); upper 2 petals nar-row spathulate, with darker markings, to 35 x 13 mm;

lower petals smaller with faint markings; hypanthium to about 10 mm; pedicel very short; fertile stamens 7.

RECOGNITION: large upright plant with thick, hairy, aromatic leaves and large flowers. *P. quercifolium* is distinguished from this species by its shorter growth habit and sticky, more deeply divided leaves.

The leaves of this species are very similar in shape to an oak but unfortunately the specific name '*quercifolium*' had already been used in 1781. It was first distinguished as a separate species in 1835, but could have been collected much earlier as it grows in river valleys on the lower slopes of mountains in southern Cape Province in a similar geographical area to *P. quercifolium*. *P. quercifolium* was introduced by F. Masson in 1774 so perhaps he collected both and they were considered varieties of one species. Hybrids between the two have been found in the wild. If allowed a free root run, it makes an impressive branching plant to over two metres tall flowering over a long period. (Plate 40)

P. papilionaceum (Linnaeus) L'Héritier, Aiton, *Hortus Kewensis*, **2**: 423 (1789)

MEANING: like a butterfly referring to flowers
SECTION: **Pelargonium**
ILLUSTRATION: Sweet **1**: 27; *Pelargoniums of southern Africa*, **1**: 32.

Erect, hairy, branched, shrub to over 2 m (6 $^1/_2$ ft) high with a strong unpleasant scent. Leaves cordate with 3–5 shallow lobes, margins crenate, to 9 × 10 cm (3 $^3/_5$ × 4 in); petiole to 8 cm (3 $^1/_5$ in); stipules broad-triangular. Flowers light to dark pink in lax-branched inflorescence held above the foliage, in clusters of 5–10; peduncles to 6 cm (2 $^2/_5$ in); upper 2 petals obovate, reflexed, pink with dark purple central mark and white basal blotch, *c.* 20 × 8 mm; lower petals very narrow, paler, *c.* 7 × 2 mm; hypanthium to 5 mm; pedicel *c.* 10 mm; fertile stamens 7.

RECOGNITION: very robust plant; flowers with very small lower petals distinguished from *P. hispidum* by less acute leaf lobes and from *P. vitifolium* by lax inflorescence.

This is one of the very early introductions having been cultivated since 1724 in the Chelsea Physic Garden. The foliage has such an unpleasant smell that it is difficult to imagine the reports that it has been used as a substitute for tobacco. However, if it is not grown in a confined area but with space in which to develop to its natural size, the flowers are a very attractive feature with showy upper petals from which the specific name is taken. In the wild, it grows in south-western and southern Cape Province on forest margins.

P. pseudoglutinosum Knuth, *Repertorium specierum novarum regni vegetabilis*, **45**: 64 (1938)

MEANING: resembling *P. glutinosum*
SECTION: **Pelargonium**
ILLUSTRATION: *Pelargoniums of southern Africa*, **3**: 113.

Decumbent or upright viscid shrub with brownish-red branches. Leaves green with dark brown purple blotch over main veins, ovate, pinnatisect with irregularly cut, acute segments, base cuneate, *c.* 3 × 2 cm (1 $^1/_5$ × $^4/_5$ in); petiole 1–2 cm ($^2/_5$–$^4/_5$ in); stipules triangular. Flowers pink; peduncle *c.* 2 cm ($^4/_5$ in) bearing 1–2 flowers; upper petals with a darker blotch and veins, oblong often with notched apex, *c*. 20 x 8 mm; lower petals slightly smaller; hypanthium to 10 mm; pedicel 2 mm; fertile stamens 7.

RECOGNITION: low-growing sticky shrub with inflorescence of one or two flowers distinguished from others with viscid foliage by the lack of scent.

This is not a widely grown species but does form an attractive low-growing shrub with relatively large flowers produced over a long season. It has a limited inland distribution in southern Cape Province. *(Plate 40)*

P. quercifolium (Linnaeus f.) L'Héritier, Aiton, *Index Kewensis*, **2**: 422 (1789).

MEANING: leaves like an oak, *Quercus* sp.
SECTION: **Pelargonium**
ILLUSTRATION: *Flowering plants of Africa*, **49**: 1923; *Pelargoniums of southern Africa*, **3**: 119.

Erect, branching viscid shrub to over 1.5 m (5 ft) in height with strong balsam scent. Leaves triangular in outline, pinnately lobed to bipinnatifid to pinnatisect, rough with long glandular hairs, usually to 5 x 5 cm (2 x 2 in); petiole *c*. 4–5 cm (1 ³/₅–2 in); stipules cordate. Flowers pale pink to darker purplish-pink, in inflorescence of 3–6; peduncle *c*. 5 cm (2 in); upper petals spathulate, apex notched, with darker purplish-pink lines and blotches, to 25 x 7 mm; lower petals smaller, unmarked; hypanthium *c*. 10 mm; pedicel 1–2 mm; fertile stamens 7.

RECOGNITION: Vigorous somewhat viscid aromatic shrub with leaves less finely divided than *P. denticulatum* and larger flowers. *P. panduriforme* has much hairier foliage and a more vigorous habit.

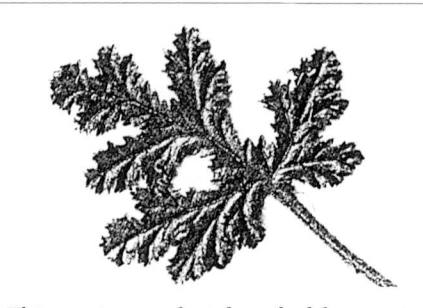

This species was first described from material collected by Thunberg, probably on his second journey accompanied by F. Masson, as it was introduced to Kew from that expedition in 1774. It is variable in leaf and flower characteristics in its native habitat on rocky slopes, at a higher altitude but similar region to *P. panduriforme*. In South Africa, plants may be found that have a great range in the degree of lobing of the leaves. Some have leaves with simple rounded lobes very similar in shape to a small oak leaf. These are close to some of the cultivars grown for many years, such as 'Royal Oak', which produce viable seed in cultivation and match many of the descriptions of the early botanists. Others have more divided leaves like those illustrated by L'Héritier as *P. quercifolium*, the great oak-leaved cranesbill. Yet others are even more divided and more like L'Héritier's second illustration of var. *pinnatifidum*, the small oak-leaved cranesbill which approaches the plant described in *Pelargoniums of southern Africa*, **3**: plate 119. It would appear therefore that at least some of the cultivars thought to be hybrids could well be geographical variants of the species. The name makes much more sense once the simple lobed plants have been seen growing wild. The local vernacular name means skunkbush, presumably on account of the strong scent of the foliage. *(Plate 40)*

P. radens H.E. Moore, *Baileya*, **3**: 22 (1955).

MEANING: refers to harsh foliage
SECTION: **Pelargonium**
ILLUSTRATION: *Botanical magazine*, **3**: 95.

Erect branched rose to lemon-scented shrub to over 1 m (40 in) in height. Leaves rather greyish-green, very rough, broad-triangular in outline but palmatisect, each section pinnatisect, the final segments barely 3 mm wide with margins rolled under, to 5 × 7 cm (2 × 2 $^4/_5$ in); petiole to *c.* 5 cm (2 in); stipules narrow triangular. Flowers pale or pinkish-purple, in a loose inflorescence of about 5; peduncle to 3 cm (1 $^1/_5$ in); upper 2 petals narrow obovate, with darker markings, *c.* 17 × 6 mm; lower petals slightly narrower, unmarked; hypanthium *c.* 6 mm; pedicel to 10 mm; fertile stamens 7.

RECOGNITION: similar to *P.* 'Graveolens' but leaf segments very narrow and the longer pedicels giving a more open inflorescence. Leaves similar in shape to *P. denticulatum* but never sticky.

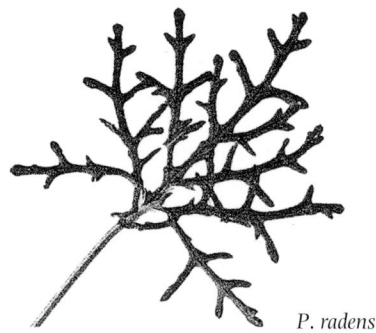

P. radens

Although this species was first introduced to Kew by F. Masson in 1774, it was known as *P. radula* until renamed *P. radens* by Moore in 1955 because of a confusion in the use of names in the later years of the eighteenth century. The true species is less common in cultivation than the plants grown nowadays as 'Radula' with rose-scented leaves, a cultivar closely related to 'Graveolens' and mentioned under that entry. *P. radens* grows wild from south-western Cape Province to Transkei. This species is considered with *P. capitatum* to be a parent of the cultivar 'Rosé' grown in great quantities for the commercial production of geranium oil on the island of Reunion. *(Plate 38)*

P. ribifolium Jacquin, *Icones plantarum rariorum*, **3:** 11, t. 538 (1792)

MEANING: leaves like *Ribes*

SECTION: **Pelargonium**

ILLUSTRATION: *Pelargoniums of southern Africa*, **2:** 120.

Erect, branching, aromatic shrub to nearly 2 m (6 $^1/_2$ ft) in height. Leaves palmately 3-lobed, divided a quarter to halfway to midrib, coarsely toothed, lobe apices rounded, base cordate, rough with glandular hairs, sometimes slightly sticky, light green, 4–5 × 5–6 cm (1 $^3/_5$–2 × 2–2 $^2/_5$ in); petiole to *c.* 4 cm (1 $^3/_5$ in); stipules triangular. Flowers white in a loose-branched inflorescence in clusters of 6–12; peduncle to 4 cm; upper 2 petals obovate, sometimes lined with dark red, 17 × 11 mm; lower petals *c.* 11 × 3 mm, narrower, sometimes lined; hypanthium to 8 mm with conspicuously swollen base; pedicel to 10 mm; fertile stamens 7.

RECOGNITION: tall aromatic shrub with white flowers in lax inflorescence.

P. 'Radula'

The first date given for the introduction of this species is in *Index Kewensis* which suggests that it was introduced by Loddiges in about 1798. However, as it had been described by Jacquin from plants cultivated at Schönbrunn in Vienna four years earlier, there must have been an earlier collector whose name was not recorded. It grows in southern and eastern Cape Province but has not become very popular in gardens of the northern hemisphere, perhaps because it only develops its full potential when given a large space in which to spread.

P. scabroide Knuth, *Repertorium specierum novarum regni vegetabilis*, 18: 293 (1922)

MEANING: rough, referring to texture of foliage
SECTION: **Pelargonium**
ILLUSTRATION: *Pelargoniums of southern Africa*, **3**: 125.

This rarely grown plant is related to *P. englerianum*. It is non-aromatic with erect or spreading reddish young stems covered with clearly visible long hairs and rather rough-textured, deeply 3-lobed leaves. The flowers are very similar in size and shape to *P. englerianum* but are white or pale pink in colour. It grows in the shade of rocks in mountainous areas of south-western Cape Province.

P. scabrum (Busman filius) L'Héritier, Aiton, *Hortus Kewensis*, **2**: 430 (1789)

MEANING: rough, referring to texture of leaves
SECTION: **Pelargonium**
ILLUSTRATION: *Pelargoniums of southern Africa*, **1**: 42.

Erect, branched, shrub to over 1 m (40 in), usually strongly lemon-scented, and covered with rough stiff hairs. Leaves variable, rough and stiff in texture, rhomboid in shape, 3-lobed, sometimes cut almost to base into narrow segments, coarsely toothed, apices acute, base cuneate, c. 4 × 4 cm (1 ^{3}/5 × 1 ^{3}/5 in); petiole about 2 cm (^{4}/5 in); stipules triangular. Flowers dark purplish-pink to almost white, in a loose-branched inflorescence in clusters of up to 6; peduncle to 2 cm (^{4}/5 in); upper petals narrow spathulate, with darker markings, to 15 × 5 mm; lower petals to 10 × 3 mm; hypanthium c. 5 mm; pedicel to 10 mm; fertile stamens 7.

RECOGNITION: lemon-scented shrub distinguished from *P. citronellum* by rougher leaves with a more rhomboidal shape and the pedicel shorter than hypanthium. Unscented plants may sometimes also be found.

The leaves of this species are variable in form and the very narrow-leaved plants have been named var. *balsameum* (Jacquin) Harvey, (*P. balsameum* Jacquin) although in the wild there is a continuous gradation between the forms. Selected distinct forms which are propagated vegetatively might be given cultivar names if this course of action is considered appropriate. However, it is not a widely grown species and, in general, cuttings are not always successful, are slow to root, and the plant itself is rather slow-growing. It grows wild in south-western and southern Cape Province and was introduced on many separate occasions. Although the first record is uncertain, the form originally described as *P. balsameum* was introduced to Kew by F. Masson in about 1790, and the type first known as *P. scabrum* introduced for the Hammersmith nursery of Lee and Kennedy in 1775. Natural hybrids with *P. lanceolatum* of the section *Glaucophyllum* and known as *P.* × *tricuspidatum* L'Hérit. show characters intermediate between the two parents. Other hybrids are known in the wild but rarely seen in cultivation. There has been confusion over the use of the specific name in some quarters. This species should not be confused with the totally unrelated, large, bright reddish-pink-flowered cultivars, 'Paton's Unique' and 'Monsieur Nonin' which are sometimes listed incorrrectly as *P. scabrum* 'Apricot'.

P. sublignosum Knuth, *Das Pflanzenreich* 4, **129**: 460 (1912)

MEANING: somewhat woody
SECTION: **Pelargonium**
ILLUSTRATION: *Pelargoniums of southern Africa*, **3**: 13.

Erect aromatic or non-aromatic shrub to about 1 m (40 in) if allowed to grow freely, usually with long hairs. Leaves dull slightly grey-green, cordate in outline, with 3 shallow lobes, margins coarsely toothed, to c. 3 x 4 cm (1 $^1/_5$ x 1 $^3/_5$ in); petiole c. 2 cm ($^4/_5$ in); stipules ovate. Flowers white to pale pink; peduncle to 4 cm (1 $^3/_5$ in) with about 7–8 flowers; upper petals obovate, with deep red-purple veins and marks, to 18 x 7 mm; lower petals narrow, unmarked; hypanthium with swollen base, 5–15 mm; pedicel 5–10 mm; fertile stamens 7.

RECOGNITION: similar to *P. englerianum* but with larger flowers and less crisped leaves.

This is known only in a few specialist collections but the scented forms should perhaps be more widely distributed as it is not difficult to propagate either by cuttings or seed. It grows in the damper habitats of the mountainous regions of south-western Cape Province. *(Plate 41)*

P. tomentosum Jacquin, Icones plantarum rariorum, **3**: 10, t. 537 (1792)

MEANING: hairy, referring to leaves

SECTION: **Pelargonium**

ILLUSTRATION: *Botanical magazine*, **15**: 518; Sweet, **1**: 168; *Pelargoniums of southern Africa*, **2**: 144.

Low-growing, but wide-spreading, branched subshrub to 50 cm (20 in) in height, strongly scented of peppermint. Leaves with 3 or 5 rounded lobes and cordate base, 4–6 x 5–7 cm (1 $^3/_5$–2 $^2/_5$ x 2–2 $^4/_5$ in), with long, soft, velvety hairs; petiole often twice length of blade; stipules triangular to nearly 2 cm ($^4/_5$ in) long. Flowers in a loosely branched inflorescence, white, sometimes faintly tinged pale purple; peduncle to 15 cm (6 in); upper 2 petals obovate, with purple lines, c. 9 x 5 mm; lower petals to 11 mm long but very narrow; hypanthium to 2 mm; pedicel to 20 mm; fertile stamens 7.

RECOGNITION: leaves softly hairy, peppermint-scented.

Sometimes known as the penny royal pelargonium because of the scent, this species is very popular as an ornamental and has also been used as a culinary flavouring and for the production of peppermint oil. The species will set abundant seed but some clones in gardens seem to be infertile. These may be hybrids or simply infertile plants, but as they resemble the species in all other characteristics they may be used in the same way and propagated vegetatively. However produced, it forms a very vigorous spreading plant with foliage both attractive for its scent and texture. It has been cultivated continually since its introduction by F. Masson in 1790, and in the Isles of Scilly it became naturalized for many years. It can be grown in semi-shaded, moist positions, similar to those of its native habitat where it grows in mountainous areas at the edges of forests of south-western Cape Province.

P. vitifolium (Linnaeus) L'Héritier, Aiton, *Hortus Kewensis*, **2**: 425 (1789)

MEANING: vine-like leaf

SECTION: **Pelargonium**

ILLUSTRATION: *Pelargoniums of southern Africa*, **1**: 49.

Erect, robust, hairy, shrub to about 1 m (40 in) in height, strongly aromatic, with a harsh, somewhat lemon-like scent. Leaves with three rounded lobes, hairy rather rough to touch, margins coarsely toothed, base cordate, to 14 × 14 cm (5 $^3/_5$ × 5 $^3/_5$ in); petiole to 10 cm (4 in); stipules broadly triangular. Flowers pink to pale purplish-pink, to 15 mm across, in a compact inflorescence of about 10 flowers; peduncle *c*. 10 cm (4 in); upper petals narrow, obovate with darker markings, often paler in centre, *c*. 15 × 4 mm; lower petals slightly narrower, unmarked; hypanthium to 5 mm; pedicel very short; fertile stamens 7.

RECOGNITION: very similar to *P. capitatum* but leaves harsh to touch and not rose-scented; plant upright.

Forms with slightly different flower colour and markings are known in cultivation and some in the past have been incorrectly labelled *P. acerifolium*; all are representative of the species. It was first recorded growing in Leiden in 1710 and at the Chelsea Physic Garden in London in 1724. It grows wild in south-western and southern Cape Province and, like many others of this section, is found among other shrubs near streams, often in mountainous regions.

Peristera de Candolle

The production of seed is important for the survival of members of this section during the dry seasons. In their native land, many of the species are pioneers of disturbed land and even in cultivation, the seed tends to germinate quite quickly. The section includes some of the more insignificant short-lived plants, with straggly weedy, often trailing habits. Usually the plants are initially quite compact, but at flowering, elongate to produce many spreading branching stems with long internodes.

The flowers themselves, borne on very fine pedicels in many-flowered heads, are small, rather regular in shape, the petals often barely longer than the sepals. *P. minimum* has no merit for its flowers, as they can be less than 5 mm across and the petals only as long as the sepals, but could be treated as a curiosity. This species is also unique for the section as it has tubers on its roots. The fertile stamens are 5 or 7, occasionally 4 as in *P. columbinum*, and the hypanthium is often very short or more or less non-existent. The leaves may be simple or divided, pinnately or palmately veined but usually with rather long petioles. In habit, several resemble species of *Erodium* or *Geranium* found wild in temperate regions of the northern hemisphere.

Many are annuals in the wild, completing their life cycle in a short space of time while water is available, but in cultivation with an adequate supply and less extreme growing conditions, they will sometimes survive for a few years, especially if the flowering stems are cut back soon after the first seeds are produced. It is a sensible precaution to allow at least some seeds to mature before pruning, to ensure that the species is not lost if it does suddenly die completely. As might be expected from this ephemeral behaviour, the pattern of rainfall is not significant in the natural distribution of the species.

This section is more extensive than any other with species which may be found in Australia, East Africa, islands such as Madagascar and Tristan da Cunha, as well as in South Africa. Many individual species themselves have a very wide geographical distribution which is unusual among most species of the genus which tend to be limited to small areas by geographical or climatic barriers. All should be grown in light soils and most may be propagated by cuttings as well as seed. Although some are interesting for the collector, few are of great horticultural value, and many are very similar in appearance.

The section was first defined by de Candolle in 1824 and named after the greek for pigeon, for the supposed resemblance of the leaves of some, such as *P. columbinum*, to a pigeon's foot. Initially it did not include those with seven fertile stamens but many were reclassified in *Flora Capensis* in 1860. One or two have been used for medicinal purposes and some have been known in European gardens for over two hundred years.

1	Stem erect, thick and almost succulent with persistent stipules	**drummondii**
	Habit not as above	2
2	Leaves simple	4
	Leaves compound or at least deeply and repeatedly divided	3
3	Plant prostrate, perennial with tubers	**minimum**
	Plant erect, usually annual	**senecioides**
4	Petals over 10 mm long	5
	Petals under 8 mm long	6
5	Flowers pale purple to violet	**iocastum**
	Flowers white or pink	**australe**
6	Flowers bicoloured, upper petals white, lower petals pink to purplish-pink	**nanum**
	Upper and lower petals similar in colour	7
7	Stamens 7, sometimes 6	**grossularioides**
	Stamens 3–5	8
8	Leaves divided almost to base, stamens 4	**columbinum**
	Leaves lobed but not divided to base, stamens 3–5	9
9	Leaves ovate with truncate or cuneate base; petals pale pink, darker on reverse	**chamaedryfolium**
	Leaves cordate; petals deep pink	**inodorum**

P. australe Willdenow, *Species plantarum*, **3**: 675 (1800)

MEANING: 'australe' means southern so the name may be taken to mean either from Australia or from the southern hemisphere

SECTION: **Peristera**

ILLUSTRATION: Sweet, **1**: 68; *Pelargoniums of southern Africa*, **3**: 12.

Straggling often hairy, sometimes faintly aromatic, herbaceous perennial with a short erect stem; the leafy flowering stems extending to about 30 cm (12 in) long. Leaves with 5–7 shallow lobes, cordate base, to 10 cm (4 in) across, hairy or more or less glabrous, sometimes flushed purple below; petiole sometimes reaching 20 cm (8 in) but usually less; stipules triangular, c. 5 mm. Flowers 5–10, sometimes more, white or pale pink, in a fairly compact inflorescence; peduncle to 10 cm (4 in); upper petals obovate with notched apex, veined red, c. 10 x 4 mm; lower petals very slightly smaller, sometimes faintly veined; hypanthium 2-8 mm; pedicel 2–10 mm; fertile stamens 7.

RECOGNITION: herbaceous plant with compact head of pink or white flowers. Stems less succulent than *P. drummondii* and foliage not rose-scented as in *P. capitatum*.

This very variable species may be found in South Australia, New South Wales, Victoria and eastern Tasmania as well as New Zealand, growing in a range of different habitats, usually in sandy dunes in coastal areas but also inland. As a result of the range of geographical locations and the variety of habitats there are many synonyms, and in gardens several quite distinct forms may be found, some of which have been given cultivar names.

In the form first identified and most commonly grown, the leaves are rather softly-hairy and the flowers white, streaked with dark red, arranged in a fairly tight head. This has been widely known as *P. glomeratum* Jacq.. This and similar forms with pink flowers show a super-ficial resemblance to *P. capitatum* with which the species has sometimes been confused. *P. australe*, however, is not rose-scented, is much smaller in both flower and leaf and never becomes woody. Another collected in Tasmania, has a rosette of dark green leaves, a reddish leaf stalk and smaller, white flowers. Yet another is similar but has attractive pink flowers which contrast with the red purple underside of the foliage and petioles. The differing relative lengths of hypanthium and pedicel can make the inflorescence of different plants quite distinct in appearance. A study of the Australian species was made in 1961[23] but it is possible that some of these variants may eventually be given specific status.

There is also confusion between this species and *P. drummondii*, and for many years all the plants received under the latter name proved to be forms of *P. australe*. However, once both have been grown for at least one season, the differences, especially in habit and flower size, become more apparent. Recently one plant was received and grown as *P. australe* 'Redondo' but, appears to be very close to *P. drummondii*.

Before flowering, the plants make rather neat rosettes but as the flower stems elongate and branch, the whole plant becomes rather untidy. Seeds are produced freely and germinate easily and it is interesting to note that the different morphological forms breed true.

Superficially this species resembles two hardy species of *Erodium* from the Mediterranean regions and these have occasionally been listed as *P. australe*. *Erodium trifolium* (syn. *E. hymenodes*) from North Africa has larger white flowers veined with red in a looser inflorescence. Its leaves have a strong unpleasant smell described by one author as resembling a rancid goat. *E. pelargoniiflorum* is very similar but with less strongly smelling foliage, sepals whose apex is extended into a sharp pointed awn and flowers held well above the foliage. *(Plate 42)*

P. chamaedryfolium Jacquin, *Icones plantarum rariorum*, **3:** 9, t. 523 (1792)

MEANING: leaf resembling *Chamaedrys*, now known as *Teucrium*

SECTION: **Peristera**

ILLUSTRATION: *Pelargoniums of southern Africa*, **2:** 30.

Low-growing, sprawling, short-lived hairy plant. Leaves hairy, ovate, pinnately lobed, sometimes deeply almost to base, margins dentate, base cuneate to shallow cordate, c. 3 x 2 cm (1 $^1/_5$ x $^4/_5$ in); petiole 2–3 cm ($^4/_5$–1 $^1/_5$ in), usually equal to lamina; stipules triangular. Flowers on peduncle to 4 cm (1 $^3/_5$ in) bearing 5–6 flowers in rather tight head; petals pink, veined dark red and darker on under side, oblong, with rounded apices, c. 7 x 2 mm; hypanthium 2–3 mm; pedicel 2–3 mm; fertile stamens 5.

RECOGNITION: small low-growing short-lived plant with tiny pink flowers, distinguished from related species by leaf shape and very short pedicel.

This is another of the species in this section with a weedy habit and tiny flowers which is only grown in specialist collections. Normally behaving as an annual, it sometimes survives for more than one year in cultivation. It grows wild in south-western Cape Province.

P. columbinum Jacquin, *Plantarum rariorum horti caesarei Schoenbrunnensis*, **2:** 4, t. 133 (1797)

MEANING: 'columbinum' refers to a dove and the species was known in the past by the common name dove's foot cranesbill, perhaps because of the leaf shape although there is not a very obvious connection

SECTION: **Peristera**

ILLUSTRATION: *Pelargoniums of southern Africa*, **3:** 30.

Sprawling plant with long visibly erect hairs. Leaves light green, cordate in outline, deeply divided, often almost to base, segments deeply and irregularly cut, about 2–3 cm ($^4/_5$–1 $^1/_5$ in) across; petiole to 6 cm (2 $^3/_5$ in), exceeding lamina; stipules ovate to 6 mm. Flowers bright to dark purple; peduncle 3–5 flowered to 4 cm (1 $^3/_5$ in); petals just longer than sepals, lined dark purple, narrow obovate with pointed apices, to 6 × 2 mm; hypanthium 1–3 mm, very prominent; pedicel 15–20 mm; fertile stamens 4.

RECOGNITION: small low-growing short-lived plant distinguished from related species by 4 stamens and deeply divided leaves.

Although easy to propagate from seed, it is not often grown, probably because of its resemblance to a rather straggly weed. In cultivation it is short-lived and dies back in hot dry conditions but not necessarily if watering is continued. It was first introduced by F. Masson in 1795 from south-western Cape Province and also was grown in Vienna when it was first described.

P. drummondii Turczaninow, *Bulletin de la société des naturalistes de Moscou*, **31**: 421 (1858)

MEANING: named for botanist James Drummond who collected many new Australian plants

SECTION: **Peristera**

Plant with semi-succulent thick stems covered with persistent brown membranous stipules, becoming more or less woody with age. Leaves dark green, veins conspicuously impressed, hairy, more or less orbicular, margin irregularly, crenate, slightly undulate, deeply cordate base with lobes usually overlapping, *c.* 4 cm (1 ^{3}/$_5$ in) across; petiole *c.* 9–13 cm (3 ^{3}/$_5$–5 ^{1}/$_5$ in), sometimes persistent; stipules broad-triangular, to 10 mm long, persistent. Flowers white to very pale pink; peduncle to 6 cm (2 ^{2}/$_5$ in) with 4–7 flowers; upper petals veined dark pink, obovate, to 14 × 12 mm; lower petals unmarked, slightly smaller; hypanthium 1–5 mm; pedicel 8–12 mm; fertile stamens 7.

RECOGNITION: somewhat similar in habit to some members of section *Cortusina* such as *P. cortusifolium* but hypanthium very short. Distinguished from *P. australe* by larger flowers and thick succulent stems.

A little known species which is very different in habit to most other species of the section and perhaps after further investigation, may be classified elsewhere. It grows in a restricted location in rocky conditions of inland areas of western Australia and was first discovered about 150 years ago. Grown by a few specialists, it can be propagated by cuttings and will also produce seed.

P. grossularioides (Linnaeus) L'Héritier, Aiton, *Hortus Kewensis*, **2**: 420 (1789)

MEANING: leaves resembling those of gooseberry

SECTION: **Peristera**

ILLUSTRATION: *Pelargoniums of southern Africa*, **1**: 20.

Spreading short-lived, almost hairless, herbaceous plant often with red tinged stems, sparingly branched and with long internodes. Leaves rounded to reniform, shallow 3–5 lobed, to 4 × 5 cm (1 ^{3}/$_5$ × 2 in), with a fruity scent similar to peach; petiole to 10 cm (4 in) or more; stipules small, triangular. Flowers many in a rather tight head, *c.* 8 mm across, almost regular in shape, usually purplish-red although pink-flowered plants may be seen; peduncle to 15 cm (6 in); upper petals oblong, to 6 × 1–2 mm, marked with deeper purple; lower petals unmarked; hypanthium *c.* 3 mm; pedicel very thin to 8 mm; fertile stamens 7.

RECOGNITION: trailing plant with fruit-scented foliage and dark purple pink flowers.

The scent of the foliage variously described as peach, fruit or coconut, makes it a popular though rather untidy short-lived plant in collections. It may be found over a wide area of southern Africa and is used in native medicine.

It is naturalized in parts of California and was cultivated as early as 1731. Some plants in cultivation under this name with variegated foliage and larger pink flowers are hybrids of *P. crispum*, and do not belong to this species. Seed is freely produced.

P. hypoleucum Turcz. (syn. *P. harveyanum* Knuth) is somewhat similar but with pale pink flowers and as the specific name suggests, the undersurface of the leaves is white hairy.

P. madagascariense Baker from Madagascar with paler flowers, and **P. apetalum** P. Taylor from Tanzania usually without petals are also related. **P. glechomoides** A. Rich. from eastern Africa is aromatic and has larger 4-petalled bright pink flowers. The one species from Tristam da Cunha, **P. acugnaticum** Thou., is similar and may prove to be a geographical variant of *P. grossularioides*.

P. inodorum Willdenow, Hortus Berolinensis, **1:** 34 (1804)

MEANING: scentless although foliage is somewhat aromatic

SECTION: **Peristera**

ILLUSTRATION: *Pelargoniums of southern Africa*, **3:** 72.

This is another species very similar to *P. grossularioides* but is found wild in south-eastern Australia, Tasmania and New Zealand. There are usually three to five fertile stamens rather than seven and the deep pink flowers tend to be even smaller with petals only slightly longer than the sepals; the foliage is only faintly aromatic. There are two other species with the reduced number of stamens: **P. helmsii** Carolin from mountainous regions of New South Wales with densely hairy obtuse sepals and **P. littorale** Hügel from coastal areas of southern and western Australia with acute sepals bearing long hairs and slightly paler pink flowers. All three species are distinct from *P. australe* by their more

P. littorale

ephemeral nature, reduced stamen number and considerably smaller flowers. A newly named species **P. renifolium** Swinbourne from South Australia has equally tiny pale pink flowers and an even more sprawling habit. None of these species has any great horticultural merit.

P. iocastum (Ecklon & Zeyher) Steudel, *Nomenclator botanicus*, **2:** 287 (1841)

MEANING: meaning unclear although 'io', greek for violet, may refer to the flower colour

SECTION: **Peristera**

ILLUSTRATION: *Pelargoniums of southern Africa*, **3:** 77.

Low-growing perennial plant branching at base from a short basal stem. Leaves often reddish below and also above as they age, cordate, shallow-lobed, occasionally deeply, apex rounded, 2–3 × 1.5–3 cm ($^4/_5$–1 $^1/_5$ × $^3/_5$–1 $^1/_5$ in); petiole to 10 cm (4 in), much exceeding lamina; stipules ovate *c.* 5 mm. Flowers pale purple, relatively large for the section; peduncle 3–4 flowered, 5–10 cm (2–4 in) held above foliage; upper petals with dark purple veins and marks, narrow obovate, with pointed appearance, *c.* 12 × 3 mm; lower petals slightly smaller, clawed, with fewer lines; hypanthium 5–8 mm; pedicel to 20 mm; fertile stamens 5.

RECOGNITION: small plant with star-shaped flowers and spreading pale purple petals.

In a pot, this makes an attractive dwarf plant when covered in flower, but the flowering season is short and it could never be described as showy.

Any seed should be retained as this plant although not annual, is short-lived. It is found in south-western Cape Province, especially at high altitudes and was not collected and described until 1835. *(Plate 43)*

P. minimum (Cavanilles) Willdenow, *Species plantarum*, 664 (1800)

MEANING: very small
SECTION: **Peristera**
ILLUSTRATION: *Pelargoniums of southern Africa*, **3**: 89.

Prostrate plant with numerous small tubers, spreading to 20 cm (8 in) or more across depending on conditions. Leaves grey-green, ovate, bipinnately divided into linear segments, to 2 cm ($^4/_5$ in) long; petiole 2–5 cm ($^4/_5$–2 in); stipules triangular, minute. Flowers white in cluster of 3–6 on peduncle to 10 mm; sepals green with white margin; petals oblong, rarely exceeding sepals, *c.* 4 x 1 mm; hypanthium usually a simple depression at base of upper sepal; pedicel *c.* 5 mm sometimes longer; fertile stamens 5 but so small as to be impossible to count without magnification.

RECOGNITION: small tuberous plant with minute white flowers.

(actual size)

Despite its minute flowers which are easily missed if not expected, it makes an interesting addition to a specialist collection with a neat habit before flowering and attractive finely divided grey-green foliage. It has been known since the late eighteenth century, and in the wild it is found in arid conditions throughout southern Africa where it is used in folk medicine. The tubers enable it to die down in very hot dry seasons but in cultivation it will flower over a long period; it seeds freely and does not necessarily die down completely. Other species such as **P. nelsonii** Burtt Davy and **P. pseudofumarioides** Knuth differ slightly, each with a distinct hypanthium and longer petals.

P. nanum L'Héritier, *Compendium generalogium*, 39 (1802)

MEANING: dwarf
SECTION: **Peristera**
ILLUSTRATION: *Pelargoniums of southern Africa*, **3**: 107.

Annual prostrate plant with branches to about 30 cm (12 in) or more depending on conditions of growth. Leaves light green, ovate, 3-lobed, margins shallow-lobed, base shallow cordate or rounded, *c.* 1.5 cm ($^3/_5$ in); petiole *c.* 1 cm ($^2/_5$ in); stipules *c.* 1–2 mm. Flowers bicoloured; peduncle to 5 cm (2 in) with 4–6 flowers; petals similar in size, upper white with darker veins, narrow obovate, 6 x 2 mm; lower pale pinkish-purple, clawed; hypanthium 1–3 mm; pedicel *c.* 5 mm; fertile stamens 5.

RECOGNITION: annual with small bicoloured flowers.

This very short-lived species seeds itself quite freely and the seed should be collected for resowing in case the tiny seedlings are accidentally weeded out. It was introduced into France, Germany and England on different occasions at the end of the eighteenth century and was grown in specialist collections. Several plants grown together create quite a colourful show when in flower but the tiny flowers do not make it a widely cultivated plant. **P. brevirostre** Dyer is somewhat similar in foliage characteristics.

P. senecioides L'Héritier, Aiton, *Hortus Kewensis*, **2**: 420 (1789)

MEANING: foliage supposed to resemble *Senecio* sp.
SECTION: **Peristera**
ILLUSTRATION: Sweet, **4**: 327; *Pelargoniums of southern Africa*, **3**: 126.

Erect branching, aromatic annual to about 50 cm (20 in) tall or more, with pubescent slightly rough stems. Leaves with somewhat thickened texture, 2 or 3 pinnatifid, *c.* 4 x 3 cm (1 $^3/_5$ x 1 $^1/_5$ in); petiole 2-5 cm ($^4/_5$–2 in); stipules triangular. Flowers white with red-purple netting on the reverse side of petals;

peduncle 2–4 flowered, 2–4 cm ($^4/_5$–1 $^3/_5$ in); upper petals with dark blotch at base, spathulate, 10–15 × 5–6 mm; lower petals without dark blotch at base, slightly narrower; hypanthium c. 7 mm; pedicel 1–2 mm; fertile stamens 5.

RECOGNITION: annual erect plant with white petals reticulated with reddish-purple on the under surface.

Found in south-western Cape Province, this species was introduced into cultivation by F. Masson in 1775. As an annual, it should be maintained by seed but it is grown more as a curiosity for its unusual-patterned petals and for its turpentine-like scent.

Polyactium de Candolle

All the plants in this section have a tuber or at least thickened roots which retain water and food enabling the plant to survive underground during periods of excessive heat or lack of rain. The tubers, many of which have been used by native people for food or for their medicinal properties, are often large, some also with additional smaller tubers, but none covered by the papery sheaves typical of members of the section *Hoarea*. The plants usually have a short thickened stem although *P. gibbosum* is unique with a scrambling almost succulent stem swollen at the nodes. The leaves are lobed or pinnately-divided and, unlike species of the *Hoarea* section are found, at least in the wild, at the same time as the flowers. In cultivation, the period of dormancy often needs to be curtailed or the leaves may die down before the flowers completely open. The almost regular flowers, each with a long hypanthium and 6 to 7 fertile stamens, are formed on a many-flowered scape.

The section divides neatly into three on both geographical and morphological grounds. Those such as *P. triste* are found in the winter rainfall area of south-western Cape Province. These species flower early in the season at the end of the rainy period and in cultivation in the northern hemisphere, before many other species of the genus. They tend to have yellowish, greenish, dark brown or purple-coloured flowers and are scented at night, presumably to attract the appropriate pollinating insects. Except for *P. gibbosum* which scrambles through bushes, the plants grow in short grassy areas and are often difficult to locate because of their dull-coloured flowers.

Plants found in the summer or all year round rainfall areas of the eastern side of the country grow in damper soils in habitats of taller grassland. They flower later in the season and all except *P. pulverulentum*, have fringed petals, but are not necessarily sweetly-scented at night.

The third group includes the very variable *P. luridum*, with a much wider distribution. It is found in many parts of southern and eastern Africa and is apparently able to survive in a wider range of habitats and climates.

The section *Polyactium* includes *P. triste*, the first pelargonium to be known in Europe, which together with many of the other early introductions, was included in the genus *Geranium*. After the genus *Pelargonium* was recognized and related species discovered, a new genus, *Polyactium*, was proposed by Sweet in 1823 for *P. multiradiatum*, to indicate the numerous radiating flowers of the inflorescence. This was relegated, like several other genera, to a section of *Pelargonium* by de Candolle in the following year.

In the past, several hybrids have been raised between species of this section but unfortunately most of these are unknown today. If the scent of the flowers could be more widely introduced into plants without a dormant stage, the results could be extremely interesting. The basic chromosome number for the section is x = 11, and the majority of the species have 22 or 44 chromosomes.

These species appear easier to cultivate than those of the section *Hoarea* even though they also have tubers and become dormant for part of the year. Although water is important during the growing period, great care must be taken to keep the plants dry during the dormant period and so avoid the chance of the tubers rotting. Most readily set seed which germinates easily but slowly. Plants grown from seed may take more than one season to flower as the tuber needs to reach a certain size. These smaller tubers may be separated and grown on. The plants flower more freely if grown in larger pots where the tubers can develop freely.

1	Petals fringed	2
	Petals not fringed	4
2	Leaf outline oblong, 3-pinnate divided into linear segments rarely over 1 mm wide	**bowkeri**
	Leaves not as above	3
3	Leaf outline ovate to oblong, shallow to deeply pinnately divided	**schizopetalum**
	Leaf outline almost circular, mature leaves deeply palmately divided into linear segments up to 5 mm wide	**caffrum**
4	Plant scrambling with succulent stems and swollen nodes; flowers greenish-yellow	**gibbosum**
	Plant usually with short stem above ground or none	5
5	Stipules linear, acuminate, stiff; leaf variable; flower colour variable from red, pink, yellow, greenish to white	**luridum**
	Stipules triangular to ovate, membranous; flowers in shades of yellow, brown or purple	6
6	Leaves glaucous, often covered with a powdery pubescence, and somewhat fleshy	**pulverulentum**
	Leaves not glaucous or fleshy	7
7	Leaves oblong, lobed sometimes deeply; softly-hairy	**lobatum**
	Leaves more deeply or repeatedly divided	8
8	Leaves 2–3 pinnate, with final segments linear	9
	Leaves usually 2-pinnate with broad flattened segments with rounded apices	**radulifolium**
9	Leaves covered with fine hairs visible to the naked eye	**triste**
	Leaves with hairs but not readily visible without magnification	10
10	Leaves with flattened segments over 2 mm wide	**multiradiatum**
	Leaves with very fine segments under 2 mm wide	**anethifolium**

P. anethifolium (Ecklon & Zeyher) Steudel, *Nomenclator botanicus*, **2**: 288 (1841)

MEANING: leaf like dill, *Anethum graveolens*
SECTION: **Polyactium**
ILLUSTRATION: *Pelargoniums of southern Africa*, **3**: 5.

This species is very similar to *P. triste* but has no hairs on the leaves visible without magnification; the petiole and leaves have a red tinge. However, from the flowers alone, it would be difficult to distinguish the two and there could be some justification for including both within one variable species. Both grow in similar situations but *P. anethifolium* has a more limited distribution in the south-western part of Cape Province. It was first described as a new species in 1835, but had perhaps been collected earlier and identified as *P. triste* or one of the several variants of this species which had been given specific status at that time.

P. bowkeri Harvey, *Flora Capensis*, **2**: 592 (1861-1862)

MEANING: named for Henry Bowker, (1822-1900), the naturalist who was the first to collect this species
SECTION: **Polyactium**
ILLUSTRATION: *Botanical magazine*, **90**: 5421; *Flowering plants of South Africa*, **17**: 671; *Pelargoniums of southern Africa* **2**: 12.

Perennial plant with an underground tuber and a very short aerial stem. Leaves grey-green, softly-hairy, narrowly oblong in outline, tripinnately-divided into soft linear segments, 20 cm (8 in) long; petiole to 12 cm (4 $^4/_5$ in); stipules narrow triangular, 10–12 mm. Flowers 5–10, pale yellowish-green, the lower petals flushed purple, *c.* 4 cm (1 $^3/_5$ in) across, all petals deeply cut into linear segments, *c.* 25 × 20 mm; peduncle leafless, exceeding the foliage, to 30 cm (12 in) in the wild but may be less in cultivation; hypanthium *c.* 20 mm; pedicel *c.* 8 mm; fertile stamens 7.

RECOGNITION: leaves feathery, grey-green and flowers with fringed petals.

Even if this plant does not flower, it is worth growing for the attractive erect, soft, grey-green feathery foliage. The first plants, illustrated in the *Botanical magazine* in 1864, were sent to W. Wilson Saunders in England in 1863 from Transkei by Cooper who collected it at the same time as Bowker. It grows wild in the grassland of eastern Cape Province and Natal, often at altitudes to 2000 m (6,500 ft), in the area of summer rainfall. The leaves are said to have been eaten by some of the local people and it has also been used as a medicinal plant. *(Plate 44)*

P. caffrum (Ecklon & Zeyher) Harvey, *Flora Capensis*, **1:** 278 (1860)

MEANING: from locality where it is found, Caffraria or land of Kaffirs

SECTION: **Polyactium**

ILLUSTRATION: *Pelargoniums of southern Africa*, **2:** 18.

Perennial plant with a large hard tuber and a short aerial stem. Leaves almost orbicular in outline, usually divided several times into linear segments, the segment margins revolute, to about 10 cm (4 in) across, sometimes almost entire or lobed; petiole exceeding leaf blade; stipules narrow triangular to 2 cm ($^4/_5$ in). Flowers many in a fairly tight head, yellowish-green to purple, *c.* 3 cm (1 $^1/_5$ in) across, all petals deeply cut into linear segments, *c.* 25 × 20 mm; peduncle leafless, to 50 cm (20 in); hypanthium 15-25 mm; pedicel *c.* 5-15 mm; fertile stamens 7.

(half actual size)

RECOGNITION: very similar in flower to *P. bowkeri* but leaves rounded with broader more rigid segments with revolute margins.

Like all three species with fringed petals, *P. caffrum* grows in grassland but is the least common in the wild being found only in the coastal areas of south-eastern Cape Province. It was first discovered in the early 1830s and is thought to have been cultivated since about 1860. *(Plate 45)*

P. gibbosum (Linnaeus) L'Héritier, Aiton, *Hortus Kewensis*, **2**: 422 (1789)

MEANING: gibbous referring to swellings at nodes
SECTION: **Polyactium**
ILLUSTRATION: Sweet **1**: 61; *Pelargoniums of southern Africa*, **1**: 17.

Spreading or scrambling almost hairless plant with succulent stems swollen at the nodes, becoming woody later. Leaves glaucous, pinnately lobed with 1 or 2 pairs of unevenly toothed or lobed leaflets at base, somewhat succulent, *c.* 12 × 7 cm (4 $^4/_5$ × 2 $^4/_5$ in); petiole about equal to leaf; stipules narrow triangular. Flowers up to 15, greenish-yellow, 1.5–2 cm ($^3/_5$–$^4/_5$ in) across, sweetly-scented at night; peduncle 5–10 cm (2–4 in); upper petals obovate, slightly reflexed, 10–12 × 6 mm, lower petals slightly smaller; hypanthium 2–3 cm ($^4/_5$–1 $^1/_5$ in); pedicel very short; fertile stamens 7.

DIAGNOSTIC FEATURE: swollen-jointed nodes on stem.

Commonly known as the gouty pelargonium on account of the swollen nodes, the stems are rather brittle but although easily broken, the pieces quickly root again if treated as cuttings. In cultivation it may need support but in the wild, along the western coast of South Africa, may often be found scrambling for several metres through shrubs. It was growing in England in 1712 but is listed in Dutch publications as early as 1687. Some of the early enthusiasts tried to use *P. gibbosum* to create new hybrids, and Robert Jenkinson produced several interesting plants illustrated by Andrews in 1805, with pink, pinkish-green or brown and yellow flowers and the habit of the species. Sweet (**3**: 213) also illustrates an unusual hybrid which he calls *P. mutabile*. This was raised from *P. gibbosum* by Colvill in 1822 with flowers which open a light purple, turning yellow-green as they age, with the typical scrambling habit and evening scent of the species. Unfortunately, none have apparently survived until today. *(Plate 46)*

P. lobatum (Burman filius) L'Héritier, Aiton, *Hortus Kewensis*, **2**: 418 (1789)

MEANING: lobed referring to leaf shape
SECTION: **Polyactium**
ILLUSTRATION: Sweet **1**: 51; *Flowering plants of Africa*, **49**: 1924; *Pelargoniums of southern Africa*, **1**: 24.

Perennial plant with a very large irregular tuber covered with rough bark and a very short aerial stem. Leaves variable in shape and degree of lobing but usually 3- or more lobed, softly-hairy, to 30 cm (12 in) long in the wild but usually smaller in cultivation; petiole to 15 cm (6 in); stipules broad ovate. Flowers sweetly night-scented in umbels of 5–20 on a branched inflorescence, very dark purple-black, *c.* 2 cm ($^4/_5$ in) across; peduncle to 30 cm (12 in) or more; all petals, margined yellowish-green, rounded, 12 × 6 mm; hypanthium to 3 cm (1 $^1/_5$ in); pedicel very short; fertile stamens 6.

RECOGNITION: large tuber, large lobed leaves and very dark, night-scented flowers.

Roots of this species were first sent to Holland from the Cape in 1698. Commelin in 1701 describes it from plants growing in the garden of medicinal plants in Amsterdam and it was in cultivation at the Chelsea Physic Garden in London in 1739, although probably growing in England earlier than this. Despite the early recognition of this species, Linnaeus considered it to be *P. triste*. It is reputed to be the parent of several primary hybrids which are still widely grown today.

P. 'Ardens' raised in 1810 has pinnately lobed leaves and scarlet flowers, each petal with a deep brownish-black central blotch, while *P.* 'Schottii' has a distinct stem, grey, hairy, more feathery foliage, and larger more purplish-red flowers, the petals marked with purple-black lines. Both are thought to be the result of a cross between *P. lobatum* and *P. fulgidum*. *P.* 'Lawrenceanum' raised about 1827 with deep purple petals margined with pale greenish-yellow and shallow-lobed leaves is probably a hybrid of *P. lobatum* and *P.* 'Ardens'.

Plants with more deeply divided leaves were once treated as a distinct species, *P. heracleifolium*, but many of these are now included in *P. radulifolium*. Many of the exciting plants illustrated by Sweet and other artists are said to be hybrids of this species although unfortunately not many are in cultivation today. Wild plants may be found in sandy conditions in south-western and southern Cape Province.

P. 'Schottii'

P. 'Ardens'

P. luridum (Andrews) Sweet, *Colvill catalogue*, **2:** (1822)

MEANING: dirty yellow, referring to colour of some flowers

SECTION: **Polyactium**

ILLUSTRATION: Sweet, **3:** 281; *Pelargoniums of southern Africa*, **1:** 26.

Stemless plant with an underground tuber. Leaves hairy, extremely variable, the early ones shallow pinnately lobed, the older, more deeply divided into linear segments as the season progresses, to 25 cm (10 in) long in the wild but rarely reaching this size in cultivation; petiole to 30 cm (12 in); stipules linear, 2–4 cm ($^4/_5$–1 $^3/_5$ in). Flowers up to 50 or more, white, pink, yellow or occasionally red, 2.5–3 cm (1–1 $^1/_5$ in) across, night-scented; peduncle to 1 m (40 in) in wild, usually less in cultivation; petals to 12 mm, broadly obovate, the lower slightly narrower; hypanthium over 3 cm (1 $^1/_5$ in); pedicel 5–10 mm; fertile stamens 7.

RECOGNITION: very variable leaf morphology, many-flowered inflorescence with large flowers in wide range of colours.

This is yet another species in the section *Polyactium* which shows considerable morphological variation resulting in numerous different synonyms over the years. As in other species, distinct vegetatively propagated forms could be assigned cultivar names if the plants were to become widely cultivated. It also has one of the widest geographic distribution of any species of the genus, being found in damp grassland regions from Tanzania to the southern coast of South Africa. First raised from seed in England at the beginning of the last century, it was collected on many other occasions. It has also been used medicinally and in concoctions for courtship rituals. *(Plate 47)*

P. multiradiatum Wendland, J.C., *Collectio plantarum*, **2**: 56 (1809)

MEANING: many-rayed, referring to inflorescence
SECTION: **Polyactium**
ILLUSTRATION: Sweet, **2**: 145; *Pelargoniums of southern Africa*, **3**: 96.

This is another of the species similar in flower to *P. triste*. It is most easily recognized by the larger leaves up to 30 cm (12 in) or more in length with much broader flattened segments. The stem above ground is much more substantial, both thicker and longer so that the plant does not have the rosette appearance of other related species. The inflorescence usually has more flowers, up to 30 on each pseudo-umbel borne on peduncles with slightly swollen nodes somewhat resembling the stems of *P. gibbosum*. There are often only 5 rather than seven fertile stamens present. It was certainly cultivated in European gardens in the early years of the nineteenth century. It was assumed to be collected from the Cape Peninsula but was not rediscovered there until about 60 years ago. *(Plate 48)*

P. pulverulentum Colvill ex Sweet, *Geraniaceae*, **3**: 218 (1824)

MEANING: powdery-white, referring to appearance of the leaves
SECTION: **Polyactium**
ILLUSTRATION: Sweet, **3**: 218; *Pelargoniums of southern Africa*, **1**: 37.

Perennial plant with large elongated tuber, smaller tubers along the roots and a short aerial stem. Leaves glaucous, hairy, slightly fleshy, almost leathery, with a white powdery appearance, shallow to deeply but irregularly cut with long hairs on margins, c. 15 x 9 cm (6 x 3 $^3/_5$ in); petiole about equal to lamina; stipules triangular. Flowers 5–12, pale yellow petals sometimes blotched dark brown to purple, 1–1.5 cm ($^2/_5$–$^3/_5$ in) across; peduncle to 50 cm (20 in), branched with small leaf-like bracts; petals, narrow spathulate, 15 x 6 mm, lower petals slightly narrower and less reflexed; hypanthium to 4 cm (1 $^3/_5$ in); pedicel very short; fertile stamens 6.

RECOGNITION: leaves glaucous, somewhat succulent and covered with a white powdery pubescence; flowers less regular than other members of section with larger upper petals, borne on leafy peduncles.

The roots of this species were considered to have extraordinary powers to resist the bullets of European soldiers during the wars of the last century and they were also used for several different medicinal purposes. Several synonyms have been used for plants collected at different times but the first named were received by Colvill, a nursery man in Chelsea, London in 1822. In flower it resembles the species of the western areas but the flattened rather jaggedly-cut leaf is distinct from all others. Its natural distribution is in summer rainfall areas in the grasslands of eastern Cape Province, Natal and Transkei.

P. radulifolium (Ecklon & Zeyher) Steudel, *Nomenclator botanicus*, **2**: 289 (1841)

MEANING: raspberry-shaped leaves
SECTION: **Polyactium**
ILLUSTRATION: *Pelargoniums of southern Africa*, **3**: 120.

At one time this species was considered to be a form of *P. lobatum*, and the first formed leaves of the two may be confused. However, as the plants develop, the leaves of *P. radulifolium* become much more divided, pinnate or bipinnatifid, and much rougher to the touch. The flowers are usually paler in colour, the petals much less rounded in shape, and there are 7 fertile stamens whereas *P. lobatum* has 6. For many years it was known as *P. heracleifolium* but like *P. lobatum*, has a wide distribution along the southern and western parts of South Africa.

P. schizopetalum Sweet, *Geraniaceae*, **3**: 232 (1824)

MEANING: split petals
SECTION: **Polyactium**
ILLUSTRATION: Sweet, **3**: 232; *Pelargoniums of southern Africa*, **1**: 43.

Geophyte with large tuber covered with brown scales and a very short aerial stem. Leaves all basal, pubescent, oblong, shallow to deeply pinnately-divided, *c.* 10 × 8 cm (4 × 3 $^{1}/_{5}$ in); petiole stout to 5 cm (2 in); stipules lanceolate. Flowers, unpleasantly scented at night, up to 20, pale yellow or yellowish-green often flushed reddish-purple, *c.* 3 cm (1 $^{1}/_{5}$ in) across; peduncle unbranched, to 50 cm (20 in) even in cultivation; all petals similar, deeply cut into linear segments, 25 × 20 mm but the lower often flushed more purple than the upper; hypanthium 3–4 cm (1 $^{1}/_{5}$–1 $^{3}/_{5}$ in); pedicel *c.* 1 cm ($^{2}/_{5}$ in); fertile stamens 7.

RECOGNITION: fringed petals, lobed hairy leaf.

The leaves of plants collected from different locations in eastern Cape Province may show considerable variation and those with leaves which are less deeply lobed tend to have more purple-flushed flowers. Some botanists have separated these into a distinct species, *P. amatymbicum*. This was collected for the nurseryman Colvill in 1821 from near Port Elizabeth, as was *P. pulverulentum*. (Plate 49)

P. triste (Linnaeus) L'Héritier, Aiton, *Hortus Kewensis*, **2**: 418 (1789)

MEANING: sad, referring to dull flower colour
SECTION: **Polyactium**
ILLUSTRATION: Sweet **1**: 85, **3**: 230 & 254 (as *P. flavum*); *Pelargoniums of southern Africa*, **1**: 46.

Geophytic plant with a large tuber as well as several smaller ones and a very short succulent aerial stem. Basal leaves hairy, somewhat carrot-like, oblong in outline, deeply divided into narrow segments to 45 × 5–15 cm (18 × 2–6 in) in the wild though usually less in cultivation; petiole to 20 cm (8 in); stipules heart-shaped. Flowers night-scented, 5–20, usually brownish-purple with broad yellowish margin, but sometimes yellow or brown, *c.* 1.5 cm ($^{3}/_{5}$ in) across; peduncle to 30 cm (12 in) or more tall bearing smaller leaves; upper petals obovate, slightly reflexed, *c.* 15 × 7 mm, lower 3 slightly smaller; hypanthium to 30 mm; pedicel 2–3 mm; fertile stamens 7.

RECOGNITION: tuberous plant with hairy carrot-like leaves and dull-brown or yellow flowers.

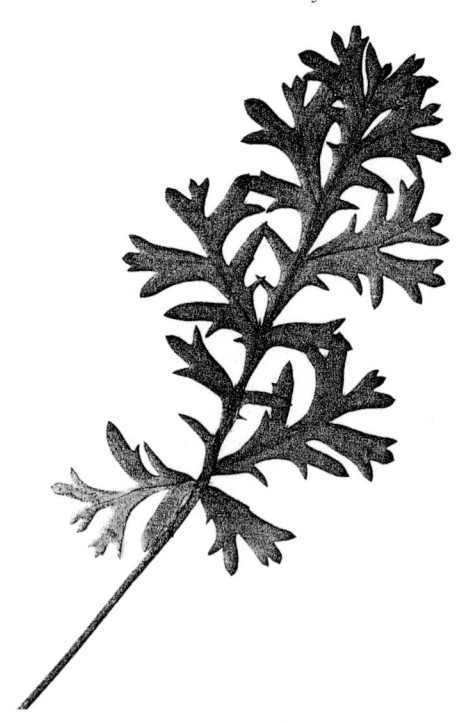

When grown from self-pollinated seed, plants may result with a range of flower colour from yellow to dark purplish-brown, and different degrees of leaf divisions. This may explain the large number of different names which have been assigned to this species in the past. *P. flavum* is a name which has been widely used for the more yellow-flowered forms of the species known since 1724, and from the horticultural point of view, there is no reason why selections which are distinct in a character such as flower colour and maintained vegetatively, should not be assigned cultivar names.

This species was the first *Pelargonium* to have been introduced into cultivation in Europe in about 1630. As it was collected by sailors on their way back from the East, it was known for almost a hundred years as the night-scented geranium from India. It may be found in sandy soils, often in colonies as a result of the spreading tubers, in the south-western and western regions of the Cape Peninsula. The tuber has been used in remedies for dysentery so perhaps this was another possible reason for its frequent collection!

The scent of this plant, and others in the section, is quite remarkable as it is emitted at almost the same time each evening and can fill a whole room with its fragrance. In the morning, no scent remains, returning only again at the same time in the evening.

Reniformia (Knuth) Dreyer

It has always been possible to divide the section *Cortusina* neatly into two groups based on their wild locality and several morphological characteristics. Following a detailed study,[11] the species found wild in the region of summer or all year round rainfall in the central and eastern parts of South Africa have now been assigned to this new section, *Reniformia*. Many of the differences in morphology reflect the fact that these do not have to survive the desert-like conditions of Namibia and the north-western part of Cape Province, experienced by those in the present *Cortusina* section. Most grow in areas receiving over 50 cm (20 in) of rain during the summer, and some have additional rain in winter.

They are mostly herbaceous or woody plants with a range of habits but do not have succulent stems nor deciduous foliage. The stipules or leaf bases are persistent and the leaves simple, lobed and frequently aromatic as in the well-known apple scented *P. odoratissimum*. The many-flowered inflorescence is branched, each individual branch bearing relatively few flowers but over quite a long season. The flowers have an irregular shape with the rather narrow upper two petals held distinctly together and erect, while the lower three are usually slightly broader and wide-spread. The hypanthium is long, the pedicel short and normally there are seven fertile stamens.

The basic chromosome number of $x = 8$ with most species having 16 or 32 fairly large chromosomes, confirms the separation of this section from *Cortusina* with small chromosomes and a basic number $x = 11$. Some species previously of the section *Ligularia*[19] have been moved into this section which is completely justified on several grounds not least, the recent creation of hybrids such as those between *P. ionidiflorum* and *P. dichondrifolium*.

Most species of this section are valued for their aromatic foliage and have been cultivated for nearly two hundred years. None are particularly difficult to grow, for as long as there is sufficient light, many will survive quite happily as a houseplant with none of the special requirements of some of the species more highly adapted to specific geographical situations. They may be rooted by cuttings or grown from seed.

1	Flowers white or pale pink	4
	Flowers bright purple-pink or deep purple	2
2	Leaves deeply pinnately-divided	**ionidiflorum**
	Leaves not pinnately-divided	3
3	Flowers almost black	**sidoides**
	Flowers bright purple-pink	**reniforme**
4	Leaves without stipules	5
	Leaves with stipules	7
5	Leaves divided almost to midrib into narrow segments	**abrotanifolium**
	Leaves not divided almost to midrib into narrow segments	6
6	Petiole at least 2-3 longer than lamina	**dichondrifolia**
	Petiole 1-2 times longer than lamina	**exstipulatum**
7	Leaf lobes angular	**album**
	Leaf lobes rounded	8
8	Plant with upright branching habit	**fragrans**
	Plant with short erect stem not branching	**odoratissimum**

P. abrotanifolium (Linnaeus f.) Jacquin, *Plantarum rariorum horti caesarei Schoenbrunnensis* **2**: 6, t. 136 (1800)

MEANING: leaves resembling southernwood, *Artemesia abrotanum*, in shape and also somewhat in scent

SECTION: **Reniformia**

ILLUSTRATION: Sweet, **4**: 351; *Pelargoniums of southern Africa*, **1**: 1.

Branching erect somewhat straggly plant to 50 cm (20 in) or more with stems becoming woody with age and bearing remains of leaf stalks. Leaves aromatic, grey-green, deeply divided more or less to the midrib into linear segments, *c*. 5–15 mm long and wide; petiole about equal to leaf or longer; stipules minute. Flowers usually white in cultivation but pink-flowered plants are also grown; peduncle to 2 cm ($^4/_5$ in), un-branched, 1–2, sometimes up to 5, flowered; upper petals narrow obovate, veined reddish-purple, *c*. 15 × 3 mm; lower petals unmarked or with faint markings; hypanthium *c*. 15 mm; pedicel 1–2 mm; fertile stamens 7.

RECOGNITION: branching woody plant differing from *P. exstipulatum* in small more finely divided aromatic leaves. *P. plurisectum* is similar in vegetative appearance but is not aromatic.

(actual size)

The typical cultivated form of *P. abrotanifolium* known for over 200 years has one or two white flowers on each peduncle and is found over quite a wide area of Cape Province, in dry often rocky situations in areas of mainly winter rainfall. However, there are other plants once known as *P. incisum* (Andr.) Willd. found towards eastern Cape Province in regions with a little rain in summer. These have up to five pink flowers on each peduncle and may have much more feathery leaves. There are very clear illustrations by several early authors such as Andrews and Jacquin showing both pink and white-flowered plants. However, these are now included in this species as in their native country there appears to be a complete gradation between the different types.[24]

In cultivation, at least three, perhaps more distinct forms are known and cultivar names have been applied to some. Investigations will be required to sort out these cultivated types and decide where cultivar names are appropriate and whether in fact the plants are forms of *P. abrotanifolium*, hybrids or perhaps even distinct species which have not been recognized. Certainly one very distinct plant is grown which is completely fertile, has feathery leaves and pink flowers. It also has very distinct stipules up to 5 mm long, a characteristic not normally associated with *P. abrotanifolium*, and this may perhaps prove to be a separate species.

P. album J.J.A. van der Walt, *South African journal of botany*, **56**: 65 (1990)

MEANING: white, referring to flowers or leaves

SECTION: **Reniformia**

ILLUSTRATION: *South African journal of botany*, **56**: 65 (1990).

Erect herbaceous plant with thick fleshy branching stems. Leaves aromatic, rather viscid with whitish tinge, palmate with 5–7 acute, sharply toothed lobes cut half way to midrib, base cordate, to 7 × 7 cm (2 $^4/_5$ × 2 $^4/_5$ in); petiole *c*. 8 cm (3 $^1/_5$ in); stipules triangular, persistent. Flowers white in branched inflorescence, each peduncle 2–4 cm ($^4/_5$–1 $^3/_5$ in) bearing about 8 flowers; upper petals sometimes with small red markings, oblanceolate *c*. 12 × 4 mm; lower petals similar to upper but spreading rather than held erect and together; hypanthium 1–2 cm ($^2/_5$–$^4/_5$ in); pedicel 5–10 mm; fertile stamens 7.

RECOGNITION: similar in flower to *P. odoratissimum* but leaves deeply and sharply cut and darker green, with erect branching stem.

This recently described species has attractive foliage with a scent which might be described as pungent or menthol-like with a hint of apple, making it a useful addition to a collection of scented-leaved pelargoniums. It may be found in the eastern Transvaal and is closely related to the more prostrate and succulent but also scented **P. mossambicense** Engl. which has white or pale pink flowers and grows wild in Mozambique as well as southern Zimbabwe and Angola.

P. dichondrifolium de Candolle, *Prodromus*, **1**: 656 (1824)

MEANING: leaves resembling *Dichondra*, a tropical species of *Convolvulaceae* naturalized in parts of Europe
SECTION: **Reniformia**
ILLUSTRATION: *Pelargoniums of southern Africa*, **3**: 44.

Perennial plant with short erect, rarely branched stem becoming woody with age and covered with remains of long petioles, to 20 cm (8 in) or more in height. Leaves crowded at top of stem, dark grey-green, with pungent scent, reniform, 1–2 cm ($^2/_5$–$^4/_5$ in) across; petiole to 10 cm (4 in); stipules minute. Flowers white; inflorescence branched, each peduncle 2–5 flowered *c.* 5 cm (2 in); upper petals oblong, with red feathered markings, 10 mm long; lower petals unmarked; hypanthium 2–3 cm ($^4/_5$–1 $^1/_5$ in); pedicel *c.* 2 mm; fertile stamens 7.

RECOGNITION: Similar in flower to *P. odoratissimum* but with much smaller darker green leaves on very long persistent petioles.

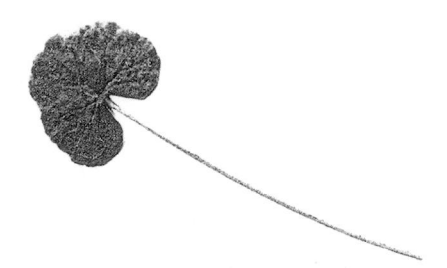

It is somewhat similar to *P. odoratissimum* but the flowering stems do not become straggly and elongated and do not bear small leaves a long their length. It was collected on several separate occasions under different names in the early part of the eighteenth century but is not widely grown despite the aromatic foliage, perhaps because of the untidy appearance of the persistent petioles.

P. exstipulatum (Cavanilles) L'Héritier, Aiton, *Hortus Kewensis*, **2**: 431 (1789)

MEANING: without stipules
SECTION: **Reniformia**
ILLUSTRATION: *Pelargoniums of southern Africa*, **2**: 60.

A branching woody shrub to about 1 m (40 in) in height though less if grown in a container. Leaves with pungent aromatic scent, viscous, grey-green, almost rhomboid in shape, irregularly toothed towards apex, *c.* 10 × 15 mm; petiole exceeding lamina; stipules minute. Flowers pale pink; inflorescence unbranched; peduncle 2–5 flowered, 2–3 cm ($^4/_5$–1 $^1/_5$ in); sepals reddish-brown; upper petals oblanceolate, with dark red feathered markings, *c.* 10 × 3 mm; lower petals unmarked; hypanthium *c.* 8 mm; pedicel *c.* 3 mm; fertile stamens 7.

RECOGNITION: small woody shrub with silvery, sticky, aromatic leaves; leaves less divided than *P. abrotanifolium*.

(actual size)

Although known to have been grown by the Countess of Strathmore as early as 1779, and illustrated in several botanical works of the late eighteenth century, and despite the attractive aromatic foliage, this species has never become a popular plant in gardens. It is easy to grow and will produce flowers intermittently over several months of the summer. It may be found wild in light soils in rocky conditions of the Little Karoo in southern Cape Province. *(Plate 50)*

P. fragrans Willdenow, *Hortus Berolinensis*, t. 77 (1806-1816)

MEANING: fragrant, referring to foliage
SECTION: **Reniformia**
ILLUSTRATION: Sweet, **2:** 172.

Small erect branching woody plant to about 30 cm (12 in) or more. Leaves grey-green with a soft velvety texture, with spicy aromatic scent, broadly ovate, crenate to lobed, 2–3 cm ($^4/_5$– 1 $^1/_5$ in) long; petiole to 4 cm (1 $^3/_5$ in); stipules 1–3 mm. Flowers white in branched inflorescence, each cluster of up to 8 flowers; sepals brownish-red or green; upper petals erect, feathered with red, oblong, to 8 x 3 mm; lower petals unmarked; hypanthium reddish-brown, to 6 mm; pedicel often brownish-red, 4–6 mm; fertile stamens 7.

RECOGNITION: small aromatic woody plant somewhat similar to but shorter than *P. exstipulatum*; leaves not viscous and flowers white. Differs from *P. odoratissimum* in its woody habit and grey-green leaves with a spicy scent.

There has always been a question about whether this plant is a true species or not and for many years, probably initiated by Sweet, this plant has been considered a hybrid between *P. odoratissimum* and *P. exstipulatum* as it shows characteristics intermediate between the two. It was first discovered in Berlin in the early nineteenth century by Willdenow who considered it to be a true species, but both putative parents had been in cultivation for many years before. It has not been found in the wild and does not readily set seed, but attempts to recreate the cross have not been totally successful. However several botanists consider that it might be a species. Other recognized species do not reliably set seed unless more than one clone is grown or the weather conditions are right. It is also possible that its native habitat no longer exists so it will never be found growing wild. Whatever the true status, it is a very popular plant in cultivation and has given rise to several named cultivars such as 'Variegatum' with creamy-yellow-edged leaves. *(Plate 51)*

P. ionidiflorum (Ecklon and Zeyher), Steudel *Nomenclator botanicus*, **2:** 287 (1841)

MEANING: violet-coloured flowers
SECTION: **Reniformia**
ILLUSTRATION: *Pelargoniums of southern Africa*, **2:** 83.

Small woody plant to about 40 cm (16 in) tall, mostly with long conspicuous glandular hairs. Leaves bright green, elliptic, pinnately lobed, pinna irregularly cut, *c.* 20 x 10 mm; petiole *c.* 10 mm; stipules minute. Flowers bright pink to magenta; inflorescence with many branches each of about 5 flowers; upper petals narrow oblong, to 15 x 3 mm, with darker lines; lower petals broader; hypanthium *c.* 3 cm (1 $^1/_5$ in); pedicel to 5 mm; fertile stamens 7.

RECOGNITION: low-growing woody plant with bright pink flowers and hairy pinnately-divided leaves.

This plant is found in the drier parts of the summer rainfall regions of eastern Cape Province adapted to both high summer and low winter temperatures. It makes an attractive plant which grows well in a hanging basket if grown indoors, and will survive outside in a mild dry climate although it is not commonly seen in cultivation. It has been used fairly recently in the USA with *P. dichondrifolium*, *P. australe*, (an inland form) and *P. odoratissimum* to create some attractive cultivars which have been named respectively 'Lavender Lad', 'Lavender Lass' and 'Lilac Lady'. *(Plate 52)*

1 *P. ovale*

2 *P. tricolor* near Garcia Pass, South Africa

3 *P. ovale* (left); *P. tricolor* (centre) and *P.* 'Splendide' (right)

4 *P. caylae*

5 *P. frutetorum*

6 *P. inquinans*

7 *P. peltatum* at Worcester Botanical Garden, South Africa

8 *P. quinquelobatum*

9 *P. stenopetalum*

10 Four variations of *P. zonale* grown from seed collected wild in South Africa

11 *P. echinatum* at Worcester Botanical Garden, South Africa

12 *P. xerophyton*

13 *P. grandiflorum* from Andrews, *The botanist's repository*

14 *P. laevigatum* near Oudtshoorn, South Africa

15 *P. lanceolatum*

16 *P. tabulare* at Kirstenbosch Botanical Gardens, South Africa

17 *P. longifolium* at Worcester Botanical Garden, South Africa

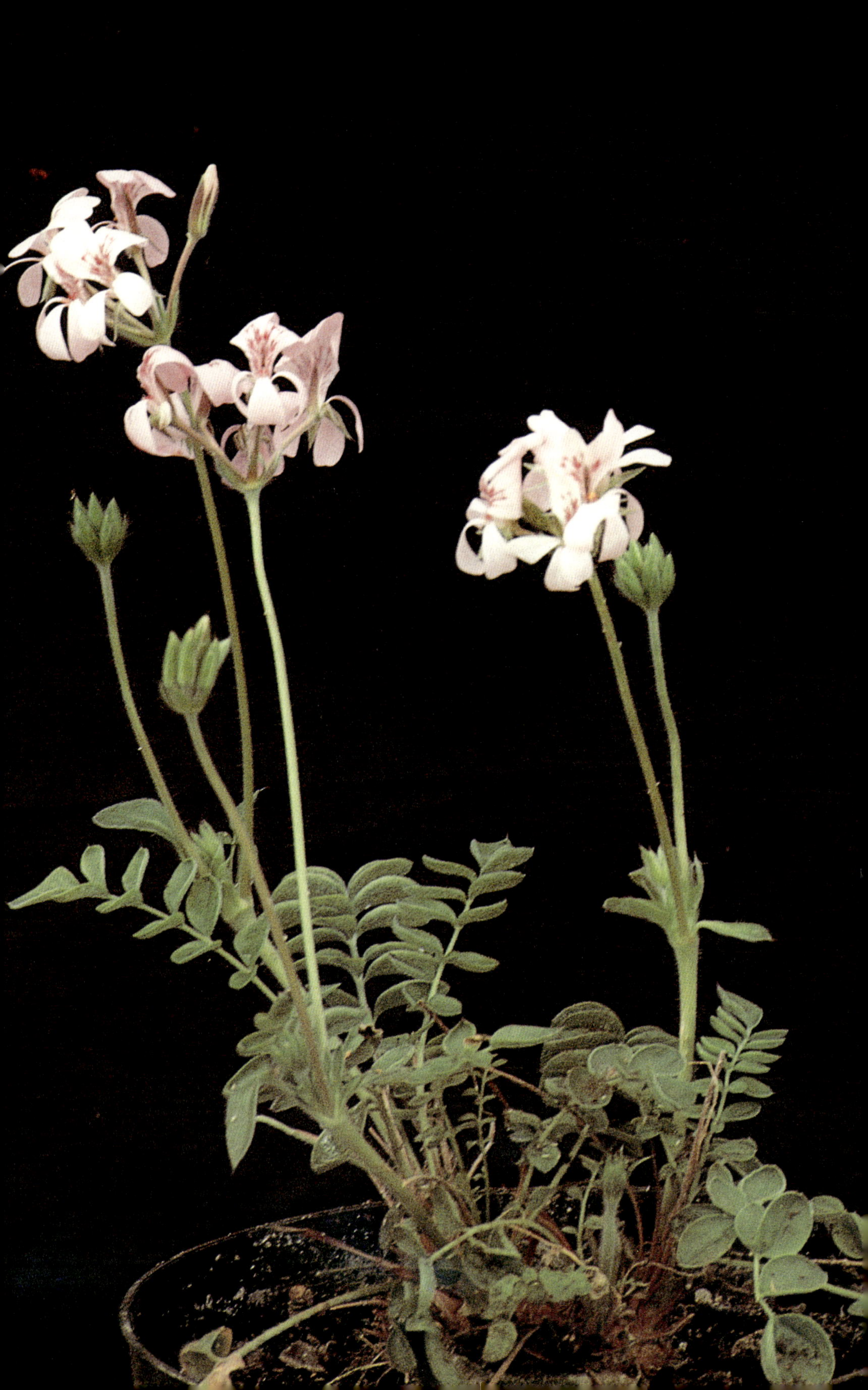

19 *P. rapaceum*

20 *P. cotyledonis*

18 *P. pinnatum* (opposite)

25 P. divisifolium

26 P. rodneyanum

27 *P. spinosum* at Worcester Botanical Garden, South Africa

28 *P. stipulaceum* at Worcester Botanical Garden, South Africa

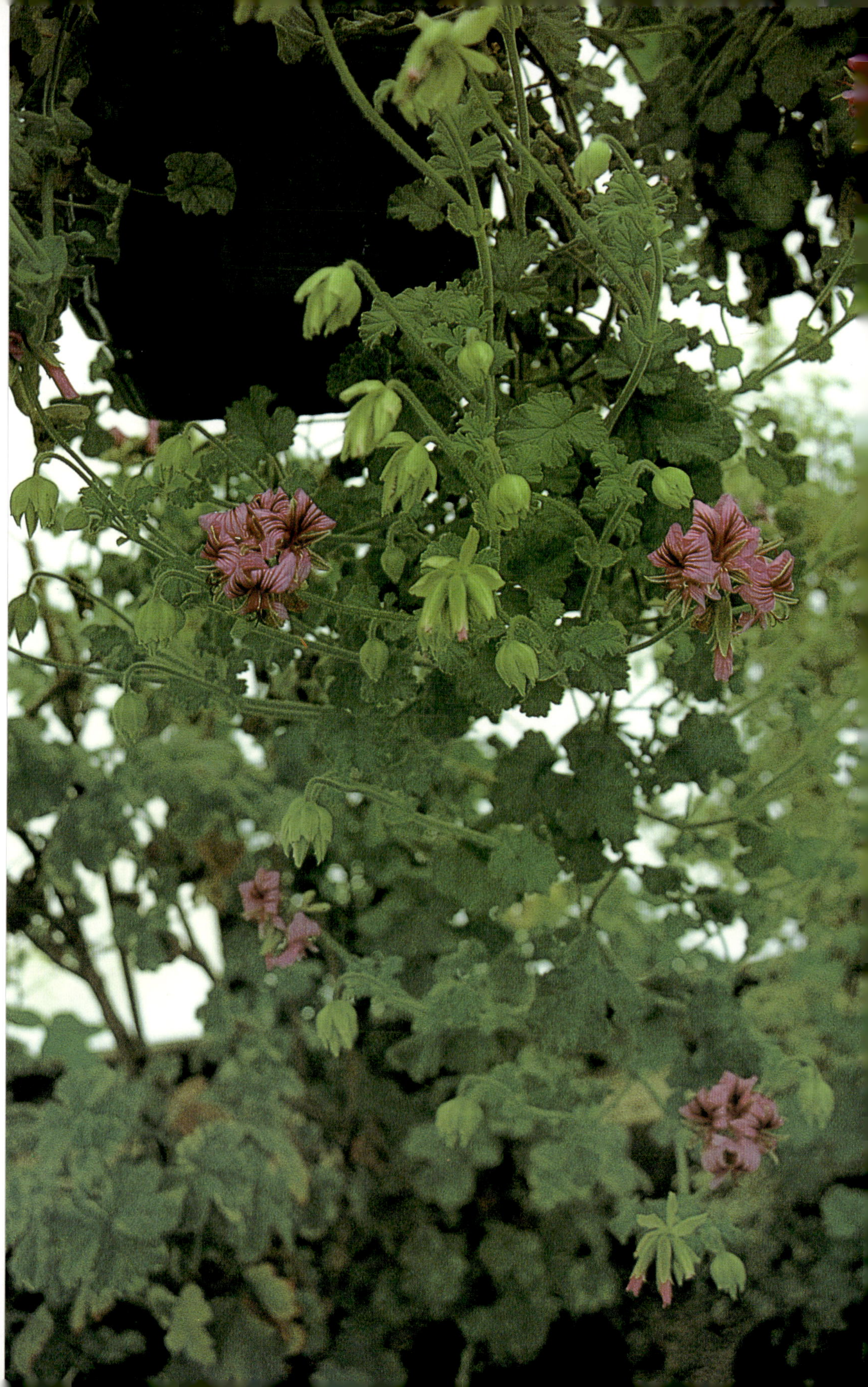

31 *P. suburbanum* subsp. *suburbanum* (opposite)

32 *P. dasyphyllum*

33 *P. betulinum*, Cape of Good Hope, South Africa

34 *P. citronellum*

35 *P. cordifolium*

36 *P. cucullatum* subsp. *cucullatum* near Rooeils, South Africa

37 *P. cucullatum* subsp. *tabulare* at Miller's Point, South Africa

38 'Graveolens' (left); *P. radens* (centre) and *P. graveolens* (right)

39 *P. hermanniifolium* from Villiersdorp, South Africa

40 *P. quercifolium* from plant collected in South Africa (top left) and grown from seed from South Africa (bottom right); *P. panduriforme* (centre); *P. pseudoglutinosum* (bottom left) and *P.* 'Royal Oak' (top right)

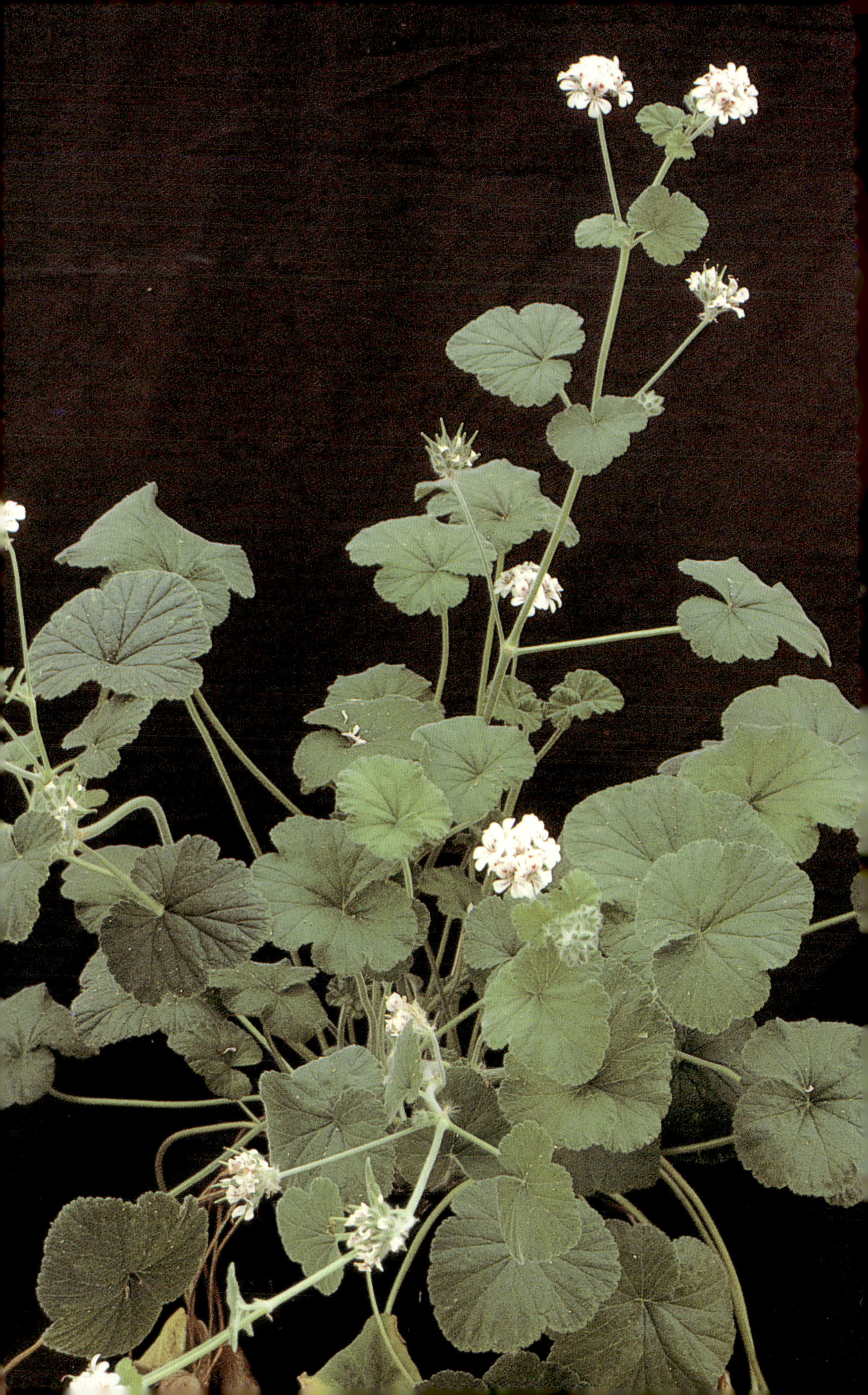

41 *P. sublignosum* (top left)

42 *P. australe* (opposite)

43 *P. iocastum* (top right)

44 *P. bowkeri* (bottom)

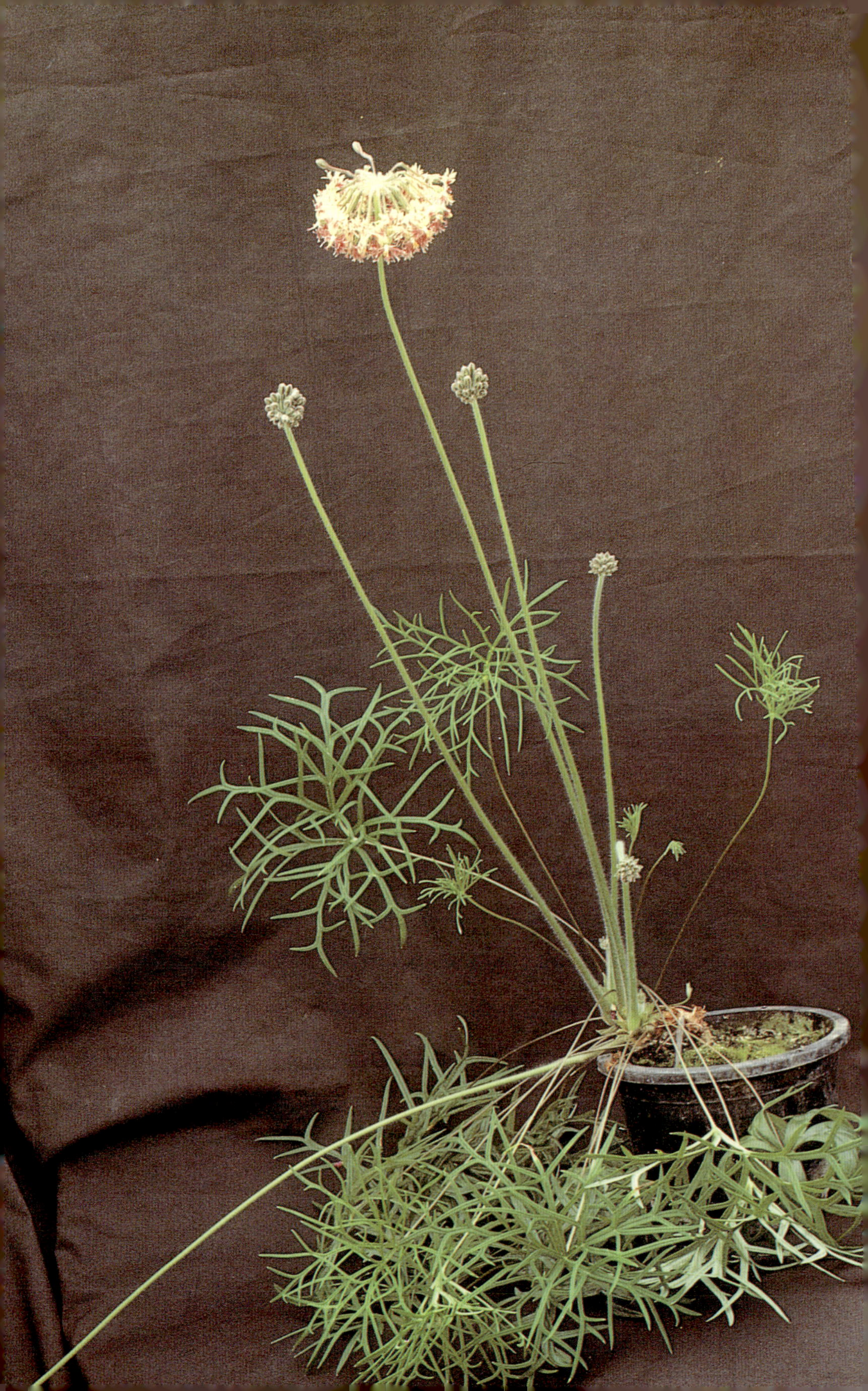

45 *P. caffrum* (opposite)   46 *P. gibbosum*

47 *P. luridum* from East Transvaal, South Africa

49 *P. schizopetalum*

48 *P. multiradiatum*

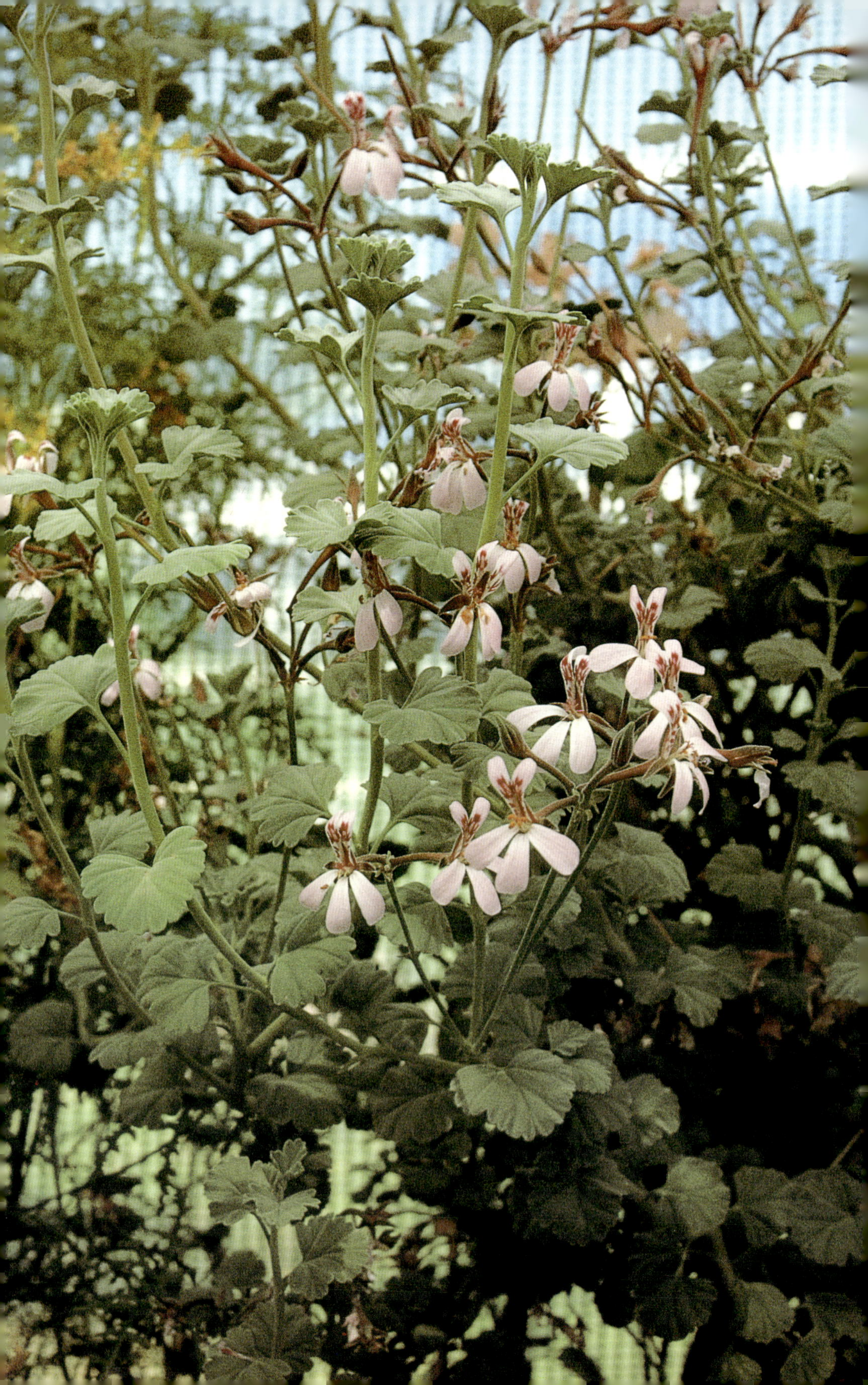

50 *P. exstipulatum* (opposite)

51 *P. fragrans* (top left) and *P. odoratissimum* (top right)

52 *P. ionidiflorum*

53 *P. reniforme* (above)

54 *P. sidoides*

P. odoratissimum (Linnaeus) L'Héritier, Aiton, *Hortus Kewensis*, **2**: 419 (1789)

MEANING: sweet-scented referring to foliage
SECTION: ***Reniformia***
ILLUSTRATION: Sweet, **3**: 299; *Pelargoniums of southern Africa*, **1**: 30.

Low-growing herbaceous plant with short thick main stem, the leaves arising from the top, but with spreading or trailing flowering stems up to 50 cm (20 in) long. Leaves light green, with a soft texture usually but not always, apple-scented, rounded with deeply cordate base, to about 4 cm (1 $^3/_5$ in) across but often much larger if growing in lush conditions; petiole *c.* 5 cm (2 in), persistent; stipules broad-triangular. Flowers white, inflorescence branched with small leaves, each peduncle 3–8 flowered; upper petals marked with red, to 10 × 3 mm oblong; lower petals slightly smaller, unmarked; hypanthium 5–7 mm; pedicel 10 mm; fertile stamens 7.

RECOGNITION: slow-growing plant with bright green, usually apple-scented leaves.

This popular species has probably been grown continually since its first introduction into Europe in 1724 where it was grown in the Chelsea Physic garden in London. In its vegetative state it forms a neat rounded plant with fresh green foliage, but the flowering branches may elongate to 30 cm (12 in) or more giving a rather straggly untidy appearance. These branches may easily be trimmed back after flowering. If grown from seed, it is possible that not all the seedlings will have such a strong apple aroma and some may have a scent which resembles the more pungent spicy scent of *P. dichondrifolium*. In the wild, it may be found in shaded places in woodland or under bushes protected from intense light over a large area in southern and eastern South Africa. *(plate 51)*

P. reniforme auct. non Curtis in *Pelargoniums of southern Africa*, **1**: 40 1977

MEANING: kidney-shaped, referring to leaves
SECTION: ***Reniformia***
ILLUSTRATION: *Flowering plants of South Africa*, **17**: 672; *Pelargoniums of southern Africa*, **1**: 40.

Erect or trailing subshrub with small tuberous roots. Leaves grey-green with velvety texture, silvery below, slightly aromatic, reniform, 2–3 cm ($^4/_5$–1 $^1/_5$ in) or more across; petiole *c.* 2 cm ($^4/_5$ in), persistent; stipules narrow triangular. Flowers bright pink to magenta; inflorescence branched, each stem to 10 flowered; upper petals narrow oblong, erect, spotted and veined darker pink *c.* 10 × 3 mm; lower petals spreading, unmarked; hypanthium 1–2 cm ($^2/_5$–$^4/_5$ in); pedicel to 10 mm, usually less; fertile stamens 7, sometimes 6.

RECOGNITION: shrubby plant with silvery-green leaves and bright pink flowers. Some forms resemble *P. sidoides* in habit but the flowers of the latter are almost black.

This species is recognizable in two distinct forms. One appears similar in habit to *P. sidoides* with a short erect stem and quite large leaves whereas the one more often seen in cultivation has more trailing woody stems and smaller leaves. Its flowers are distinct in colour and also the petals are broader and less recurved. It flowers towards the end of the summer in cultivation in the northern hemisphere and has been known since its first introduction by F. Masson to Kew in 1791. The tuberous roots have been used as a traditional medicinal remedy both for livestock and humans. It may be found in dry sandy or grassland areas of southern and eastern Cape Province. Plants have been collected in the most western part of its native habitat which are similar in flower colour and form to this species and similar in leaf shape but not texture to *P. sidoides*. It is possible that this will prove to be a new species once it has been investigated fully.

There appears to be some confusion over the identity of *P. reniforme*. The authority for the name is usually given as Curtis from the description and illustration in *Botanical magazine* plate 493 (1800). Sweet describes and illustrates a very similar plant (**1**: 48). However, examination of the illustrations show plants which resemble *P. magenteum* J.J.A. van der Walt illustrated (as *P. rhodanthum* Schltr.) in *Pelargoniums of southern Africa*, **1**: 41 (1977). Sweet likens his plant to *P. echinatum* and *P. cortusifolium* but by no stretch of the imagination could this be so if he were referring to the species known by this name today and illustrated in *Pelargoniums of southern Africa*, **1**: 40.

There are many references by Sweet and others to hybrids between *P. reniforme* and *P. echinatum*. This would appear quite feasible if the plant to which they refer was in fact what is known today as *P. magenteum* and therefore in the same section, *Cortusina*, as *P. echinatum*. The plant *P. reniforme* of today is unrelated with a different basic chromosome number, and hybrids would appear unlikely. For the time being, the name *P. reniforme* is retained for the plant described above though it would appear that it really merits a new name, and that *P. magenteum* might more correctly be known as *P. reniforme*. This could cause quite a lot of confusion among growers of the two plants! *(Plate 53)*

P. sidoides de Candolle, *Prodromus*, **1**: 680 (1824)

MEANING: resembling *Sida* sp., referring to leaf shape

SECTION: **Reniformia**

ILLUSTRATION: *Pelargoniums of southern Africa*, **3**: 131.

Plant with thick underground roots and short erect stem, covered with persistent petioles and stipules, from which arises a rosette of leaves. Leaves grey-green with silvery sheen, slightly aromatic, shortly-hairy, heart-shaped, toothed, to 4 cm (1 $^3/_5$ in) across; petiole 10 cm (4 in); stipules triangular to 10 mm. Flowers deep blackish-purple; inflorescence with many branches, each up to 8-flowered held well above the foliage; upper petals narrow oblong, often twisted and curled backwards, 12 x 3 mm; lower petals similar but slightly broader; hypanthium 25 mm; pedicel 1–2 mm; fertile stamens 7.

RECOGNITION: plant with silvery-green leaves and deep purple-black flowers.

Although closely related to *P. reniforme* sensu J.J.A. van der Walt, this species often grows in more grassland areas but will adapt to a range of habitats and altitudes from the Transvaal to southern Cape Province. It was collected by the Swedish botanist Thunberg in 1772 and also by F. Masson but never became popular in cultivation. Like many other species, it is used for its medicinal properties. *(Plate 53)*

Cultivation

The art of the horticulturist is to reproduce conditions as closely as possible to the natural habitat and climate of any particular plant so that it will thrive in the artificial situations of cultivation. However, each species has become adapted by natural selection over thousands of years to a particular microclimate and it is unrealistic for a gardener to mimic every unique set of conditions.

Fortunately plants are very adaptable and by looking at the different habitats, the general basic necessities are easily recognized. In the wild, plants have to compete for space, light and water, they have to resist diseases and browsing animals, whether large or minute, and overcome seasonal changes of temperature or rainfall. The surviving plants therefore become tough. If a gardener nurtures rare or expensive plants and cossets them in garden conditions, the resulting plants are often more vigorous but also softer than their wild counterparts. With pelargoniums this is especially so and it may prove initially a little difficult to equate a wild plant with a perfectly grown cultivated specimen free of any defect, since it has been protected from pests, diseases, drought and so on. This does not mean that growing the plants favourably is incorrect, nor does it mean that the cultivated plants are not true species.

Growers may have different ideals. For botanical research, perhaps the plants should be treated more harshly. However, for the average enthusiast the aim is to produce an attractive plant and in any case, the grower may not be able to reobtain a rare species in which case it might be wise to take a few precautions by growing it carefully. If the plants are grown simply for their ornamental value, it will not matter how lush they are or how unnatural they appear, nor is there any necessity to balance the needs of botanical exactitude and aesthetic value. There has therefore to be a compromise between the natural and artificial as it would be impractical for each individual of a mixed collection of species to be given an exact duplication of its wild environment.

Plants grown in containers are already in an artificial environment and watering especially is more critical since the roots restricted by the pot cannot elongate to seek available supplies of water together with the nutrients dissolved in it. The majority of pelargoniums grow in areas of relatively low, usually seasonal, rainfall and have become adapted to periods of drought. The soils are usually free-draining and many grow in semi-desert conditions with little vegetation, so there is also little water-retaining humus produced from the decay of other vegetation. The light is intense with long hours of sunshine. Some species experience snow and frost for a short period, but in gardens the humidity must also be low for cultivated plants to be able to survive such low temperatures.

The general factors to consider are light, temperature, water, including the humidity of the atmosphere, and nutrients, but all are interdependent. For example plants can survive lower temperatures if the humidity is low and watering is reduced. Raising the temperature in winter will reduce the humidity but may also encourage the plants to grow, even though the light intensity may be too low and they may then succumb to disease. Regular observation of the plants will quickly indicate if the conditions need to be modified. Individual cultivation techniques for plants of different sections may be found in the relevant chapters but more general factors are included here.

Light

The intensity of the light where the majority of the species grow in South Africa, is noticeably higher than in Europe and many parts of North America. Light is essential to all plant growth but even more so to the *Pelargonium* species. A few grow at forest margins, in the shade of rocks or in long grass, and if necessary these may be grown under other taller plants or certain individuals may be slightly shaded.

The majority grow in full sun and do not thrive in low-light levels. Except in very poor summers, this is rarely a problem, especially if the plants can be put out into the open, but in winter, it is important that the glass of the greenhouse is clean. Specialized horticultural lamps are available if conditions are exceptionally dull. Cuttings will, however, need extra lighting to root successfully in autumn or early spring. If shading is considered necessary, because of excessively hot conditions in a greenhouse, shading paints, blinds, slats or netting may be employed but these should not impede any ventilators and should be removed well before winter begins.

Plants grown indoors or even in a shaded conservatory often suffer from lack of light and must be placed in the brightest position possible. In winter, if the temperature is relatively low and watering kept to a minimum, light is less important. However, if the plants are encouraged to continue growth by watering and increasing the temperature, additional artificial light may become necessary.

Water

Overwatering is perhaps one of the major causes of failure in the cultivation of *Pelargonium* species. All grow in areas of limited rainfall and are adapted to overcome this problem in a variety of ways. Those of the sections Pelargonium and Ciconium tend to grow in damper situations, and are slightly more tolerant of excess water which is perhaps why the parents of all the popular cultivars belong to these sections. Conditions for others may be slightly more critical which may be why they have not remained so common in cultivation, but few are really difficult to grow.

Those native to the deserts and semi-deserts like *P. crassicaule* should be allowed to dry out completely during their dormant period. Even when growing quickly they do not require very much water. For the majority, it is advisable to allow the soil to more or less dry out before rewatering. If a peat-based compost completely dries out, it is sometimes difficult to wet the compost again by watering from above and the pot may need to stand in a shallow water-filled tray until the peat has reabsorbed the water. Be careful not to leave the pot standing in water for any length of time. As a general rule, it is better to underwater than overwater, and to reduce the amount as the temperatures drop and the light intensity falls, until the following spring when growth becomes more active. The use of capillary matting or an equivalent may be useful in a greenhouse in summer if it is not possible, for whatever reason, to water regularly, but it is not ideal for a mixed collection of species.

Watering by hand is much safer as the water requirement of each species is quite different and ideally each plant should be watered individually for its own needs. Yellowing leaves are often a sign of overwatering and the only cure is to dry out the compost and remove the discoloured leaves allowing new foliage to appear. Succulent species are best treated as any other cactus or succulent. The geophytes must not be watered at all during their dormant phase or the tubers will rot, but when the first shoots appear water may be applied sparingly until the leaves or flowers are fully developed.

Humidity

Some water vapour in the air is necessary but in a cool damp atmosphere pelargoniums become prone to disease. Little can be done to remedy this in the open but in greenhouses, the atmosphere can be modified. In summer this is only a problem in cool wet seasons and as long as there is good ventilation and any dead leaves or faded flowers are quickly removed, there should not be a problem. If the heat is excessive and the air very dry, it might be helpful to damp down the paths although this is more important for the highly-bred zonal and regal cultivars than for the species. In cold but damp winters, raising the temperature will reduce the humidity, but to reduce the prevalence of disease, ventilation is even more important. Except in temperatures near or below freezing, some vents should be kept open to ensure air circulation. Spacing the plants will also increase air circulation.

Temperature

The only hardy pelargoniums are those from the Middle East, *P. endlicherianum* and *P. quercetorum* and even these need very careful watering and very well-drained soils to survive a few degrees of frost. All the other species should be treated as frost-tender despite the fact that some experience some frost and snow in their native land. Many forms of greenhouse heating are available but it is worth being aware of the fact that some do produce water vapour, thus increasing the humidity. In winter, dry air and light are more important than high temperatures and a temperature of about 5 °C (41 °F) is normally sufficient if other factors are suitable. Except when temperatures below zero are expected, ventilation to ensure air circulation even in mid-winter is essential. A few species are slightly more tender needing a temperature nearer 7–8 °C (44–46 °F) but these are indicated individually. As long as a house is well-ventilated, high temperatures in summer are less critical unless the plants are in very small pots which dry out quickly causing the plant to become stressed. If temperatures regularly exceed about 25 °C (77 °F) shading might be considered, but it must be removed before winter when light intensities fall. In very hot weather, if the paths have to be sprayed to increase the humidity, this is best done during the middle part of the day so there is not a high degree of humidity at night as temperatures fall. For cuttings a slightly higher temperature is helpful.

Soil and nutrients

Basically there is a choice between two types of compost, soil-based or soilless; most people have their own personal preference. Each has its advantages but *Pelargonium* species will grow in either as long as it is well-aerated. They will not, however, tolerate waterlogging which is one of the quickest ways of killing them. The quality of a soil-based compost depends on the quality of the soil used in the mix, so the composition is less easy to standardize. The nutrients will also last for longer after repotting, so feeding becomes less vital. In a soilless compost, feeding is easier to regulate and many commercial growers find it more convenient. A soilless compost may be bought ready-prepared but it is not difficult to experiment with the preparation of an individual recipe to suit the specific requirement of the plants in cultivation. Most are based on peat but other substances such as coir fibre are now available and could be tried. Peat will retain more water than is suitable for *Pelargonium* species so the addition of an aerating material such as horticultural, (not builders') sand, grit, or vermiculite is essential for their health. For the succulent and geophytic types this can be increased to almost equal quantities. A pH of about 6.8 is ideal but the majority of species are fairly tolerant.

Whatever compost is used, nutrients will eventually be necessary but the majority of *Pelargonium* species, except for the very vigorous ones, need far less than the zonal and regal cultivars. A liquid fertilizer may be added when watering and one recommended for tomatoes appears to work well. It is worth remembering that in general, a high nitrogen content encourages foliage and potassium will aid flowering. Alternatively slow release fertilizers, which will release nutrients for up to nine months, may be incorporated into the compost.

Although the use of fertilizers will delay the need for repotting, the plants will eventually need a larger pot, or at least a change of some of the compost which will have become stale. This should be done with care to avoid excessive root disturbance, preferably not while the plant is in flower. Geophytic species and others that have a period of dormancy are best repotted at this time. Usually the new pot will be slightly larger but large plants, which need new compost but not enormous pots, and those whose roots have not completely filled the existing pot, may be returned to a similar sized pot after some of the old compost has been replaced with new. After repotting, however carefully this is done, the plants will suffer some stress and some leaves may turn yellow and drop. This can be partially avoided by using moist compost, not watering, and keeping the plants slightly shaded for a few days afterwards.

Clay pots are perhaps marginally better for pelargoniums as excess water will evaporate from the surface. They are heavier and therefore more stable but also more expensive. Plastic pots are lighter, easier to handle and clean, easier to store, cheaper to buy and retain water for slightly longer.

Pruning

The decision to prune species of *Pelargonium* will depend on why the collection is being grown as well as on the individual species. The geophytic and annual species die down each season and will never require pruning. The succulent species will eventually become very large but pruning will destroy the shape which may take several years to regain, so pruning should only become necessary if space is very limited. If a collection is being grown to study the natural habit of the plants, and if space is unlimited, the ideal would be to allow the plants to grow unchecked. *P. cucullatum* and *P. panduriforme* make impressive shrubs but in a greenhouse eventually reach the roof and therefore must be trimmed back so that they do not cast too much shade and so their shoots do not touch the glass and become frosted in winter.

A balance must be struck between retaining the natural habit of the species and the practicalities of growing in a limited space. The more vigorous pelargoniums can also become rather ungainly so judicious pruning will produce a more aesthetically pleasing plant. Some species are naturally bushy plants and need little attention, whereas others like *P. inquinans* can be encouraged by pruning to become less lanky. Yet others such as *P. salmoneum* can never be made to produce neat or compact plants. Again, the gardener must decide whether to adapt the situation to suit the habit of the plant or prune to change its natural shape.

Pruning is a matter of common sense. A clean sharp knife or secateurs should always be used to avoid damage or infection to the stems, and the shoot should be cut at a slanting angle immediately above a node. This will ensure that there is no die back of the stem above the cut and the buds at the node can develop. The removal of any dead or dying branches is essential for the health of the plant and unhealthy stems should be cut back to healthy tissue. Dead or dying branches should not be confused with the petioles and stipules of many species which are naturally persistent. It is not necessary to remove these. The long trailing flowering stems of species such as *P. mutans* and many members of the sections Peristera and Myrrhidium, die back naturally after the seeds have ripened, so trimming these will tidy up the plants. By cutting these stems back and continuing to water, some of the shorter-lived species may be prevented from dying completely.

Large vigorous species may be cut back to a manageable size during the growing season and any material may be used as cuttings. However, hard pruning is not advisable in the spring just before flowering as the flower may be lost for an entire season. Pruning from late autumn to early spring should also be avoided as it may allow diseases to enter into the cut surfaces and flourish in the cooler damper conditions when the plant is not actively growing. During the tidying up of plants at the end of the summer, the thinning out of many of the larger plants will encourage air circulation, reducing the leaf area and so the risk of damage by mould during the cooler damper season.

Where and how to grow and use pelargonium species

The variation of form and habit of the species makes them suitable for a range of conditions. The majority will be grown in containers in a greenhouse or conservatory. They are suitable as house-plants but need the sunniest place possible to avoid the leaves becoming yellow and the plant etiolated and eventually dying. If they become too leggy, it is worth trimming them back and returning them to the lighter conditions of a greenhouse or conservatory to recover.

There are many other possible situations where pelargoniums can be grown. Most benefit from being planted out in the open ground where they are able to become large plants. If they are grown in countries where protection is needed over winter, they may be too large to repot in autumn, but cuttings may be taken in the summer to provide smaller plants, more manageable for the winter, ready for the following year. The smaller species may be placed in a rock garden and the larger in borders. The trailing species like members of the section Myrrhidium, adapt naturally to hanging baskets giving a far better display than when grown either in a standard pot or in the ground, and an additional bonus is the extra space created in a small greenhouse. The more vigorous *P. peltatum* may be utilized like any other ivy-leaved cultivar. Window boxes, tubs and ornamental pots also make appropriate containers.

The succulents make suitable companions for other cacti and succulents as they will thrive in similar conditions. Most geophytic members of the sections Polyactium and Hoarea could be grown in an alpine house. They are dormant for many months of the year so while they are in flower they may be brought into the house or greenhouse and given a space where they may be kept dry until growth recommences.

The enormous variety of scents of the many aromatic species such as the peppermint-scented *P. tomentosum* make them suitable for planting in a herb garden, and they would be especially useful in gardens for the blind where the texture as well as the scent of the foliage could be appreciated. The foliage of many is used for flavouring. None are poisonous but the leaves should be washed thoroughly and any fungicidal or insecticidal sprays should not be employed immediately before use. Leaves of *P.* 'Radula' or perhaps one of the other rose-scented relatives of *P.* 'Graveolens' impart the traditional 'geranium' flavour to cakes and sponges and may also be added to apple jelly. Lemon flavours are provided by *P. crispum* and apple by *P. odoratissimum*. It is worth experimenting with the different scents and flavours which are useful in both sweet and savoury dishes. They may even be used as teas, to provide bath oil or in creams. The leaves may be picked and dried for use in winter, or in pot-pourri and herb pillows.

7

Propagation

In the wild, by far the most prevalent means of reproduction is by seed. In a few cases, pelargoniums may propagate themselves asexually by means of tubers becoming separated from the parent plant, by spreading underground stems or roots, or by the natural layering of branches, but all species produce seed which is the normal means by which they are both reproduced and dispersed in the wild. The horticulturist is fortunate to have the opportunity to choose from a range of methods to suit the plant and situation. The advantage of a vegetative method of propagation is that all the resulting plants will be identical to the parent and even if no seed is formed, duplicate plants may be produced. It is also possible to maintain a particularly good form of a species using this method but on the negative side asexual means of reproduction remove the possibility of natural diversity within a single species.

Vegetatively, a new plant is produced quickly and will also generally, flower quite soon after it has become established. By taking reasonable care, the majority of cuttings will produce new plants and there can be no possible confusion about their identity. If the plant appears to be failing for whatever reason, vegetative methods may be the only way to save it, but care should be taken not to propagate the disease at the same time. However, a plant may be very slow-growing or its shape may be ruined by taking cuttings so vegetative means are not always the answer.

Propagation by seed is slower and more erratic but is eventually a much more satisfying method. Plants resulting from seed, whether in the wild or in cultivation and without any hybridization, show some degree of variation and by the process of natural selection, those most adapted to any one particular environment will flourish. In this way, a species will be successful in survival and as each plant has an efficient method of seed dispersal, the species is able to colonize other geographical areas. The seedlings most well adapted to the slight differences in conditions will survive.

Pelargoniums show a great degree of diversity even within any one species and it is this natural variation in the morphological characteristics which sometimes makes them so frustrating to identify. This is quite apart from the fact that in the pampered conditions of cultivation the same plant may look very different to its wild relatives. Seed raised species are therefore not necessarily identical to the parent and an interesting new variation may be found, but this does not mean that it is a hybrid.

There may be concern that seeds collected from plants growing with other species may be hybrids. Hybridization certainly does occur but far less frequently than might be imagined. There are several reasons for this lack of natural hybridization. Although all species of *Pelargonium* are sufficiently closely related to be included within the same genus, they are assigned to different sections based not only on their morphological characteristics but also on other less visible features such as their chromosome numbers. Even artificially, it often proves discouraging and frequently unsuccessful to try to cross species from different sections, although crosses between species within the same section are more common. This factor immediately reduces the number of possible hybrids in a mixed collection. Differences in flowering season prevent cross-pollination in other cases. The flower structure of the various species shows a considerable range of size and shape which in the wild is adapted for the natural pollinating insect. In particular, the length of the hypanthium is very variable so that insects with different mouth parts will be effective in the pollination of specific species but not necessarily all. It will also be observed that at any one time, insects tend to collect nectar from a limited number of plants rather than randomly from a wide range, even within a mixed collection.

Sometimes it is even a problem to artificially self-pollinate a species as some appear to be self-sterile. This may be because in cultivation, only one clone is being grown. However if plants of the same species are obtained from a different source and used for pollination, fertile seed may result. It is also true that there are ideal conditions for the production of pollen and if the weather is cool or damp, especially early in the flowering season, the pollen can be infertile or the anthers not fully formed. All these factors tend to limit the amount of cross-pollination that occurs in cultivation but if the production of pure seed is important, bees and other possible pollinators may be kept out of a greenhouse by using a fine mesh over doors and vents.

Seed grown plants may not flower in their first season as they appear to require a certain period of time to reach maturity, so to satisfy ones natural impatience, this may be overcome by taking cuttings of the young seed raised plants which will then often flower, depending of course on the time of year, very soon after they have become established and much earlier than the parent plant from which the cutting was taken.

Stem cuttings

The easiest way to propagate a plant vegetatively is by stem cuttings. This is possible for the majority of *Pelargonium* species, and cuttings should be taken during spring and summer when the light is at its most intense and the atmosphere less likely to be humid. They can be taken at almost any other time of the year but may not be successful unless extra light and heat are provided. Choose a healthy vigorous non-flowering shoot, not one that has become old and woody, and preferably one in which the internodes are not elongated. The ideal length is approximately 8 cm (3 $^1/_5$ in) long, depending of course on the species concerned, with about 3 nodes and a healthy growing tip. A cut should be made with a sharp knife immediately below the node, avoiding any damage to the stem, and any lower leaves and stipules which would be below the soil surface once the cutting is inserted, should be removed. If the remaining leaves are large, it is advisable to reduce excess transpiration by cutting them in half. The cutting is inserted into a suitable compost, either soilless or a soil-based one such as John Innes No. 1 without excess nutrients but with the addition of roughly equal quantities of an aerating substance such as horticultural sand, grit or vermiculite. The compost should be moist but not waterlogged and lightly firmed. The pots, labelled with the name of the plant, date and any other necessary details, need to be kept at a temperature of about 10–12 °C (50–53 °F) in a well-ventilated, well-lit situation but away from direct sunlight or draughts.

Several precautions will ensure success. Perhaps the most important is the use of clean tools, washed pots and sterile compost to avoid any infection which might rot the cuttings before they are able to root. Hormone rooting powder is rarely necessary but the application of a fungicide in solution watered into the compost or as a powder dusted onto the base of the cutting may be helpful especially in cool humid conditions which encourage grey mould. At the same time as taking the cuttings, the parent plant should be checked to ensure that there is no stem left above a node where the cutting has been taken, as this might die back. With experience a rooted cutting is easily recognized by its firm new growth but if in doubt, the plant may be carefully lifted from the compost. To avoid too much of a check in their growth, it is advisable to pot up cuttings into the selected medium as soon as the roots have appeared but keep them out of direct sunlight until they have re-established themselves. If a propagator with bottom heat is utilized, the cover is best removed as it may cause too much condensation which would be harmful to the cutting.

Division

The majority of pelargoniums do not naturally lend themselves to division. Some such as forms of *P. australe* produce clumps which may be split into smaller portions and repotted into a suitable compost in smaller containers.

Root division and cuttings

Several of the geophytic species such as *P. pinnatum* form not one but several tubers which may be used for propagation. Smaller tubers may be separated from the roots at the beginning or end of the dormant season and potted into individual pots. They should be planted just below the surface of the compost and as sufficient heat and water are supplied, will come into growth to produce new plants. These tuberous species are even more sensitive to overwatering and a free-draining compost, which is kept moist but not wet, is even more essential than for many other species. It must be remembered that the tubers are also sensitive to frost, and if they are not to be repotted immediately, must be kept in a frost-free environment. Plants with large tubers such as *P. lobatum* may be cut into smaller sections and repotted as long as each part has a bud. It is advisable to dust the cut surfaces with a fungicide to avoid the possibility of rotting and again, a waterlogged compost will be fatal.

A few species have very short stems so that it is impossible to make stem cuttings, and no tubers, so that division of the tuber is not another option. If seed is also unavailable and if there is no other alternative, it is sometimes possible to take root cuttings. This method will need the thickest roots, which are placed in the compost vertically with the upper end level with the soil surface. Except in summer, these may need bottom heat to encourage the development of shoots and roots. A few species will naturally send out shoots from their roots which may be used for repotting.

Layering

This is not a method of propagation usually associated with pelargoniums but may be tried with species such as *P. peltatum* which naturally spread along the ground and sometimes root on their own.

Commercial methods

More sophisticated methods of propagation such as the use of meristem culture are available to the commercial grower. This is a very useful means of producing a large number of identical plants quickly. It is also possible to use only the very tip of the growing apex and in this way produce virus-free plants from a virus infected plant. It is however impractical for the amateur as it requires sterile conditions and the use of a series of carefully formulated media containing the correct concentration of plant hormones and nutrients for each stage of growth.

Seed

When seed is ripe, it is naturally released as the mericarps break away from the fruit, and in the wild, the coiled, feathered awn will carry them some distance from the parent. In cultivation, the seed may be collected as soon as the seeds twist upwards before they separate completely, and should be stored in dry envelopes or containers and labelled immediately since it is very easy to collect seeds of several different species and forget which is which. A firm plump seed is almost always fertile. Seed may be sown in a seed compost mixed with an equal quantity of horticultural sand or an inert substance such as vermiculite or grit to increase the aeration of the medium. The compost should be moist but never waterlogged and the seed just covered. A layer of grit or sand on the surface is often helpful. Larger seeds may be pushed individually into the compost leaving the awn exposed.

The pots should be labelled with the name of the species, date, and if necessary the source of the seed, left in a frost-free situation and watered only if the compost becomes dry. The two rather thick ovate or rounded seed leaves emerge first and are followed by more typical foliage, but even these first few true leaves are not necessarily truly characteristic of the species. After the first pair of true leaves have opened, the seedling may be potted on into a well-aerated compost. Holding the young plant gently by the leaves will avoid damage to the stem. The seedlings should be kept out of direct sun for a few days in a well-ventilated position and potted on and given fertilizer when appropriate. Some people have found that germinating seed on damp blotting paper and potting up as soon as the root appears is effective. Others recommend the use of a fungicide sprayed onto the soil as a precaution against the damping-off of the seedlings.

Fresh seed is more likely to germinate within a few weeks but the germination time of *Pelargonium* seed is notoriously erratic and often may take up to two years or more, so much patience is needed and the temptation to throw out the seed pots must be resisted. Seeds have an inbuilt mechanism which ensures that germination occurs at the best time for growth when water and temperature especially are correct, otherwise the resulting seedlings might succumb to cold, heat or drought. In the wild these mechanisms are essential for the survival of the species and seed may lie dormant for many years. Many pelargoniums are colonizers of land which has been burnt by fires which have swept through their native habitats. The fire weakens the seed coat allowing sprouting as soon as rain falls. However, even when conditions appear to be ideal, germination is often erratic as if each seed has a time clock, so that some germinate quickly whereas others must await their turn several weeks later. This is yet another trick of the plant to aid survival for if all the seeds germinated together but the rain was less than expected, all would die with none left for the future continuation of the species.

Spacing the seeds sometimes results in a more even germination as the seed coat contains an inhibiting agent which prevents neighbouring seeds from germinating, thus avoiding overcrowding so that many plants die. Sometimes the removal of the seed coat encourages faster germination, or the nicking of the pointed end of the seed may allow moisture to enter more easily. In most cases, however, germination will continue to be rather slow and erratic whatever method is tried. Recently, work carried out in South Africa has revealed that chemicals within smoke help to break the dormancy of some seeds. A product has been produced by the National Botanical Institute at Kirstenbosch which results in better germination of the seeds of many South African species including Pelargoniums.

Hybridization

The attempt to create totally new plants by the hybridization of species and the repeated crossing of the resulting offspring, has produced the thousands of cultivars grown today; but the possible combination of characters is endless and much enjoyment may be gained by experimenting. To avoid disappointment, study the relationships of the proposed parents remembering that the more closely they are related, the greater the chance of success. Within a section, many crosses are possible if the chromosome numbers are similar or at least have the same basic number, and crosses between related sections are sometimes possible.

To affect a deliberate cross the pollen of one parent must be transferred to the ripe stigmas of the second without the influence of its own or stray pollen from elsewhere. The anthers first dehisce, and then the five stigmas, which until they are receptive are closed together, open outwards into five lobes. At this point the ripe pollen from the selected male parent may be dusted with a fine clean paintbrush from the anthers onto the receptive stigmatic surfaces. This may be repeated several times to ensure pollination has occurred. Although the anthers and pollen in an individual flower ripen at different times, flowers in the same head will be at different stages of development so precautions must be taken to avoid pollination by flowers in the same inflorescence. The anthers of the potential female flower should be removed before their pollen is shed, and the flower covered before the stigmas are ripe.

Once pollinated, the flower should be isolated by enclosing it in a muslin or paper bag. The pollinated flowers must be labelled at once with the date and the name of the male parent. Within a week or so, the seed head will begin to develop and the typical long beak-like fruit, from which the genus received its name, will begin to elongate. The seed will take about six weeks or so to become fully mature and should be collected as soon as the mericarps begin to split. Fertile seeds are firm and may be sown straight away. It can be difficult to remember all the crosses made on each plant, however good ones memory, so labelling and keeping records is essential. Sometimes the resulting plants may be disappointing but self-pollination of the first generation of seedlings may result in new and even more exciting variations. Although the continual perseverance by dedicated growers has produced the range of shapes and colours seen today, the original species will always hold more attraction for many enthusiasts.

8

Pests, diseases and disorders

Although the majority of *Pelargonium* species are not prone to pests or diseases, care must be taken to inspect plants regularly and any problems treated as soon as they occur. Any new plant, from whatever source, should be inspected carefully and if there is any question about its health, be quarantined before it is added to an existing collection. Healthy plants are less susceptible to pests and diseases and good cultivation techniques will help to avoid many sources of potential trouble. Simple precautions such as the removal of all debris, dead or decaying plant material, and other unnecessary rubbish such as broken pots, seed packets and so on will remove sources of infection and hiding places for some pests.

Weeds under the staging of the greenhouse may also encourage certain pests and diseases and should be removed by hand. Spraying with weed killer in a confined space is not advisable in case the plants of the collection are themselves damaged. An annual autumn clean out of the greenhouse, washing the whole structure with a solution of Jeyes fluid, is a prudent habit to foster. Any residual insect pests and fungal or bacterial spores lurking in the crevices will be eliminated, thus reducing the chance of an infestation in the following season. Cleaning any build-up of algae on the glass, especially where the panes overlap, will also increase the light intensity so essential to the *Pelargonium* species through the gloomy winter months.

Strong plants may be developed by creating the ideal conditions for growth, with bright light and a well-aerated compost being essential for *Pelargonium* species. Overwatering will encourage a stagnant compost which pelargoniums do not tolerate, and excess humidity will encourage grey mould especially if there is poor air circulation. Pelargoniums, with few exceptions, are frost-tender, but do not require very high temperatures so in winter, unless frost is expected the greenhouse may be kept open to ensure good air circulation. These conditions will also help to avoid any physiological disorders which might arise.

Sick or pest-ridden plants should be treated, destroyed or removed from other plants and not used for propagation unless precautions are taken to avoid contamination. Equipment such as knives used for propagation or pruning, should be kept scrupulously clean to avoid cross-infection of any latent disease. Pots should be cleaned thoroughly before use and compost sterilized if possible.

If treatment with fungicides or insecticides becomes necessary, it is important to follow the manufacturers instructions exactly. Many are not only lethal to the pest or disease, but also harmful to humans. Increasing the recommended dose rarely has any beneficial effect and will often cause damage to the plant being treated. The time gap between treatments for insect pests has been carefully analyzed by the manufacturer to obtain the maximum effect on the life cycle of the insect, so it is sensible to follow these directions. Some of these chemicals rely on contact with the pest or disease and these are known as 'contact insecticides' or 'fungicides'. For these it is important that all parts of the plant are covered by the treatment. Careless spraying which does not cover all of the plant may give only sub-lethal doses to the insects or fungal disease, and the survivors can build up a resistance to the chemical, so creating generations of populations more resistant to future applications of the spray. Other chemicals are absorbed into the sap of the plant remaining effective for a long time, so that sucking insects are killed when they feed many days after treatment. These are known as 'systemic insecticides'.

In persistent cases of infection, different products may be tried until the most effective cure is found. It may be worth alternating treatments on a regular basis to avoid the problem of a build-up in resistance to any one substance. As well as sprays, smoke cones may be used. These can be compared to a firework which produces a smoke of chemical rather than a colourful flash or an explosion. They are extremely effective in reaching all parts of an enclosed space by vaporizing the chemical through the air, but are obviously of no use in the open. Plants in bright sunlight appear to be more sensitive to certain chemicals, and some species of *Pelargonium* such as *P. scabrum* are prone to damage, so as a precaution, any chemical treatment is best carried out in dull conditions and if possible other methods tried first. It should be remembered that any chemical may harm useful insects such as ladybirds and hoverflies as well as the unwanted pest or disease.

Biological control, which is when a natural predator of the pest is used for its control, is the most natural method. This technique needs careful management of the environment to create conditions ideal for the reproduction of the predator. Supplies of suitable predators are available from specialist firms who will include detailed instructions for their use. The predators usually need to be reintroduced each season because if they have done their job effectively, all the pests will have been eliminated and the predator will die out by starvation. Also winter conditions are rarely suitable for the predator's survival. The timing is crucial for two reasons. Firstly, the predator must be introduced in time for it to reproduce in sufficient numbers to control the pest. On the other hand, there must already be some pests on the plants by the time the predators emerge or are introduced otherwise they will be unable to survive through lack of food. Any applications of chemicals should be withheld for some time before predators are introduced since they cannot be used at the same time. The supplier will be able to give advice on this subject.

Most enthusiasts of the species of *Pelargonium* grow a wide range of plants from all sections of the genus, so diseases and pests are less likely to take hold because of the variety of plants with different degrees of susceptibility to different problems. Even persistent pests such as whitefly have their preferences and are easier to control in a varied collection. However, in a greenhouse full of zonal cultivars, an infection of rust, for example, may spread quickly through the whole house.

PESTS

APHIDS

There are many types of aphid some black, some green, some grey. For much of the time they are wingless insects breeding very quickly, but at certain times of the year some individuals become winged and will spread to other areas. Greenfly are the type of aphid most frequently found on pelargoniums but fortunately most *Pelargonium* species are not their first choice of food. Aphids attack the young shoots and leaves, weakening the plants, distorting the young leaves and shoots and excreting honey dew on which sooty mould may grow causing unsightly black deposits. More seriously, especially in their winged phase, the insects can transmit some of the serious incurable virus diseases.

Aphids may be controlled by a variety of insecticides, both systemic and contact. Natural plant extracts such as pyrethrum or derris may also be tried. If natural predators such as the larvae of ladybirds, hoverflies, lacewings and so on are to be encouraged, avoid any chemicals which may kill these insects. Root aphids may be controlled using commercial preparations which are sprayed onto the surface of the compost.

BEES

Bees might be considered a pest by the few who do not want any cross-pollination between the species but any serious hybridizer will have taken other precautions to avoid the creation of unknown hybrids. He will in any case destroy any seedlings of unknown parentage.

CATERPILLARS

The larvae of some moths feed on the flower buds or the foliage of the larger-leaved pelargoniums. Some may be found in apparently folded leaves which have been webbed together by the insect for protection. Unless there is a major infestation, caterpillars can usually be controlled by picking them off by hand.

PELARGONIUM BUTTERFLY

In southern Europe, the adults lay eggs in the flowers and the resulting larvae eat their way through the stems of the plant causing it to collapse without much warning. By this time, treatment is too late to save the plant. This is a pest of the fleshy-stemmed zonal types of *Pelargonium* but is less important in a mixed collection of species. In cooler climates it is killed off in winter and therefore is naturally restricted in its spread.

MEALY BUGS

Although this pest is often found on house and greenhouse-plants, it is not usually a serious problem with *Pelargonium* species except that it does sometimes attack the more succulent types. It may be recognized as a dirty white insect about 5 mm long, which secretes a white hairy covering protecting it from contact chemical sprays. Systemic insecticides may help to eradicate this pest or if there are only a few, dabbing each individual with a cotton swab dipped in methylated spirit should remove them. These are sucking insects which are most likely to be found in the leaf axils and stem branches and secrete honey dew on which sooty mould may thrive.

Root mealy bugs

Poor growth or wilting might indicate that this pest is present in the soil. It is recognized as small white insects about 2 mm long or by a powdery wax-like substance with which the insect surrounds itself. An infestation may be controlled using a liquid insecticide applied to the soil. Unfortunately the mealy bug thrives in the drier soil preferred by pelargoniums but a complete drench to the soil will make it too wet for the plants. If the infestation is serious, new cuttings in fresh compost may be the best remedy.

Red spider mites

There are not a major problem with *Pelargonium* species but the effects are usually recognized before the minute wingless mites are detected. The leaves may appear yellowish and dull and the underside may be brownish or speckled and covered by a fine web. The mites are encouraged by hot dry weather so under these conditions, raising the humidity of the atmosphere by spraying the greenhouse floor and bench at midday will help to control the problem. Biological control may be tried using the red spider mite predator but it must be reintroduced into the greenhouse each season and the conditions must be right to encourage its rapid establishment. Alternatively a range of chemicals may be tested but the red spider mite has become resistant to many. If an assortment of different species is being cultivated and the plants are healthy and growing vigorously, there should rarely be cause for concern except with zonal and ivy-leaved cultivars.

Sciarids

Small black flies about 3 mm long seen on the surface of the compost are probably sciarids. In the adult stage they cause no damage. However, the larvae hatching from eggs laid in the soil live on decaying plant material and may enter the stems below ground level and cause some damage to the tissue of weakened plants. They are more prevalent in soilless and peat-based composts.

Slugs and snails

These two pests may sometimes cause damage to young shoots and fleshy leaves and can be controlled by using commercial bait. They should not be a serious problem in a clean, tidy greenhouse without excess debris or pots under which they may hide.

Vine weevils

Vine weevils are an old pest which can affect any plant and once introduced into a greenhouse can cause much damage. The adult is recognized as a brownish-black insect about 1 cm ($2/5$ in) long, which lays its eggs in soft soil. These hatch into large fat white larvae which will devour all roots and tubers, causing the plants to collapse. This pest has become much more of a problem in recent years because of the ban on the chemicals which were the most efficient means of control. There are now commercial products available but these are not yet freely available to the amateur.

Whitefly

This is probably the most common pest to be found on pelargoniums and once established, an infestation is not easy to eliminate as the insects breed very quickly, especially in the summer months. If the leaves are moved, the tiny white flies of about 3 mm long will fly up from the plants. The flattened, wingless, pale greenish-white, young stage of the insect will be seen on the under surface of the leaves. All stages suck the sap, both weakening the plants and transmitting diseases.

Like several other sap sucking insects, they excrete the sugary honey dew which encourages the unsightly sooty mould.

Many insecticides are available to the amateur but it is worth trying several as strains of whitefly have become resistant to the different brands. Yellow sticky traps are thought to attract whitefly and may be hung in the greenhouse. A very dilute solution of washing-up liquid may be sprayed onto the whitefly which breaks down their waxy protective coating so they are dislodged and become more sensitive to chemical sprays. A parasitic wasp is available to the amateur which may be used as a biological control. These are bought from specialist suppliers as eggs and if the greenhouse conditions are correct, the adult wasps will emerge and quickly begin to breed. It is wise, however, to avoid a heavy infestation by constant vigilance in the early part of the year and by squashing any adult whiteflies, or removing leaves bearing the young stages. Also plants such as tomatoes or fuchsias, which attract whitefly, should not be grown in the same vicinity.

WOODLICE

Woodlice are rarely a problem to *Pelargonium* species, causing only minor damage, and may be easily deterred by cleanliness and the removal of debris.

DISEASES

BACTERIAL BLIGHT

This fungal disease is a serious problem in zonal and ivy-leaved cultivars especially in America but fortunately less so in Europe. Affected plants show spots on their leaves and eventually die. There is no cure and the only solution is to burn the plants and sterilize the whole area. Plants brought from a reliable source should not be affected. Fortunately, the majority of species are not susceptible but care should be taken with members of the section Ciconium, from which the zonal and ivy-leaved cultivars have arisen.

BLACK LEG

Cuttings are most at risk from black leg but weak or sickly mature plants can also be affected. A black mark at the base of the stem may be the first sign but this spreads quickly upwards causing the death of the whole plant. It is not a major problem with species and if the compost is not too wet may be avoided completely. If observed early enough, cuttings may be taken from the tip of the plant before the disease reaches that part. Most pelargonium stems become black when they die from whatever cause, so not all black stems signify black leg.

GALLS

Sometimes, strange tumour-like growths appear around the base of the stem just above or below ground level. These may weaken the plant slightly but do not appear to cause significant problems. It has been thought that these are the result of a soil-borne bacteria and in the past it has been recommended that no cuttings be taken from the plants and that all plants infected should be burnt. However, in a specialized collection, these may be the only plants in cultivation and there is no strong evidence to indicate that the cause of the gall or tumour is transmitted to the resultant plants through the cuttings. The condition does not appear to be easily transmitted to nearby plants and the galls themselves may be broken-off and destroyed. Some of these galls appear to be due to an imbalance of artificial fertilizers in the compost or the result of the use of a rooting hormone.

GREY MOULD

This disease is most likely to become a problem in autumn and winter if the temperature is too low, the ventilation insufficient or the humidity too high. It may be recognized as a grey fungal covering on decaying foliage and may attack young seedlings or cuttings and eventually kill a weak plant. All dead leaves and flowers should be removed regularly, especially during the vulnerable times of the year and any infected parts of the plant cut away and the material destroyed. The knife should not then be used on healthy material to avoid spreading the spores. The ventilation should be increased so that there are no pockets of stagnant air. If necessary, the heating may be increased. However, gas and paraffin heaters often produce more moisture in the atmosphere so this does not always solve the problem. In severe cases, chemical treatment may be necessary but unfortunately many strains of the mould have become resistant to the chemicals which are available to the amateur grower. Cultivation techniques and careful monitoring of the environment are therefore even more important. Except in extremely cool damp summers, the disease is rarely seen during the growing season.

PELARGONIUM RUST

This disease is more of a problem of zonal cultivars and is occasionally seen in their ancestral species such as *P. inquinans*. It is recognized by yellow spots or rings on the upper surface of the leaves and by rust-brown patches on the corresponding part of the under side. The spores are spread by wind currents and by contact. Infected plants should not be introduced into a collection; any infected leaves should be removed and destroyed. The remaining plants may be sprayed with an appropriate fungicide and the treatment repeated at the intervals recommended by the manufacturer.

SOOTY MOULD

Black deposits on the surface of leaves are the result of a mould growing on the honey dew excreted by insects such as aphids. It detracts from the appearance of the plants but unless it is so thick as to obscure the light received by the foliage, does not cause any damage. It is not easy to remove but may be gradually washed away by rain or by watering. Destruction of the pests causing the problem is the only treatment.

VIRUS

Viruses are normally transmitted by sap-sucking insects such as aphids or whitefly but may also be passed on from an infected plant when taking cuttings. Although virus-infected plants cannot be cured, they are not always visibly affected by the disease. Some viruses cause distortion of the leaves or streaking of the petals and in a few cultivars the virus that causes this effect is encouraged, but in a collection of species, the infected plants are best destroyed. If healthy stock is purchased and the level of sap-sucking insects controlled, severe viral infections should be easy to avoid. Pelargoniums are known to harbour the black tomato ring spot virus without apparent harm to the plant, but it appears to affect tomato crops. For this reason, the import and export of pelargoniums from some parts of the world is restricted without appropriate health certificates.

PHYSIOLOGICAL DISORDERS

FASCIATION

This rarely occurs in species but is occasionally seen in the highly-bred cultivars of zonal or regal pelargoniums. The cause is uncertain but may be the result of insect attack or some other adverse condition causing damage to the growing tip at a critical stage of development.

OEDEMA

The ivy-leaved pelargoniums, sometimes also the zonals, are most likely to exhibit oedema, and *P. peltatum*, the species from which the ivy-leaved cultivars are derived, may sometimes show this condition. It is manifest as raised pale brown, corky bumps on the undersurface of the leaves and stems. These are thought to be caused by the plant taking up too much water too quickly, perhaps because it was over-dry initially or had been severely pruned and then overwatered. It is a problem that will rectify itself. However, damaged leaves may be removed if they are too unsightly, and as a precaution ensure the area is well-ventilated and the plants not waterlogged.

WILTING

This may indicate that the plant needs water but first check the compost as overwatering is a more common cause of sick *Pelargonium* species than underwatering. It may indicate that the root system is faulty or the stem has been attacked by a pest or a bacterial or fungal infection causing stem rot.

YELLOW OR RED LEAVES

It is normal for old leaves to turn yellow, brown or sometimes reddish before falling, and therefore coloured leaves do not necessarily indicate that there is a problem. However, in some cases it may be an indication that the plant has suffered a check or that the watering regime is incorrect. The presence of many red leaves usually indicates that the plant has suffered a sudden drop in temperature. This is only a temporary effect which will gradually disappear and may be avoided in plants put outside in summer by a slower hardening-off period after the protected environment of a greenhouse.

Yellowing leaves usually indicate incorrect watering. Both underwatering and overwatering may cause the same effect. Many *Pelargonium* species are particularly sensitive to overwatering and waterlogging is a frequent cause of yellowing leaves. The check in growth caused by repotting may have the same effect and a few leaves may become yellow and fall. Whatever the cause, it is important to remove the yellowed leaves, especially in the cooler, more humid autumn days to avoid grey mould. Occasionally yellowing leaves may be a sign of nitrogen shortage but this is rarely the case in the species.

POOR FLOWERING

This may be the result of insufficient light or heat at the right time of year. It may also indicate insufficient potash; the use of an appropriate fertilizer may help.

GLOSSARY

APICES

1. acuminate
2. acute
3. obtuse

BASES

4. cuneate
5. truncate
6. cordate
7. auriculate
8. peltate

MARGINS

9. entire
10. lobed
11. crenate
12. dentate
13. serrate
14. irregularly toothed

SHAPES & VEINING

15. pinnate simple
16. palmate simple
17. pinnate compound
18. palmate compound

Glossary (1)

Glossary 151

SHAPES

19. linear
20. lanceolate
21. oblanceolate
22. spathulate
23. oblong
24. elliptic
25. ovate
26. obovate
27. reniform
28. cordate
29. pinnatifid
30. pinnatisect
31. palmatifid
32. palmatisect
33. trifid

Glossary (2)

Acuminate tapering to a long narrow point (glossary fig. 1)
Acute sharply-pointed (glossary fig. 2)
Annual a plant which grows, flowers and dies within one season
Anther the part of the stamen bearing the pollen (fig. 7)
Apex (pl. **apices**) the end of the leaf, petal or stem away from the point of attachment
Auricle (adj. **auriculate**) a lobe, usually in pairs, at the base of a leaf or petal (glossary fig. 7)
Base the end of the leaf, petal or stem at the point of attachment
Awl-shaped short, very narrowly triangular structure with pointed apex
Awn a thin, stiff bristle often found on tips of leaves or sepals
Binomial the Latin name of a species consisting of two words; the first name is the name of the genus, the second, the name of the species. The binomial system was introduced by Linnaeus
Bract a modified leaf associated with the inflorescence (fig. 6)
Calyx (pl. **calices**) the outermost whorl of flower parts, see also sepals (fig. 7)
Chromosomes the thread-like bodies found in the cells of a plant, bearing the genes which are responsible for the inheritance of the characteristics of an organism. The chromosome number of a species is normally constant and is often similar for related species
Ciliate fringed with hairs as for example on the margin of a leave
Claw the narrow base of a petal
Clone a group of individuals which have arisen by vegetative reproduction from one ancestor
Columnar with a narrow, erect habit

Compound a leaf composed of several leaflets (glossary figs. 17 & 18), or an inflorescence with several branches
Cordate heart-shaped (glossary figs. 6 & 28). This may refer either to the outline shape of a leaf, stipules or petals or to the base of a leaf
Corolla the collection of petals (fig. 7)
Crenate of a leaf margin with rounded teeth (scalloped) (glossary fig. 11)
Cultivar a cultivated plant, the result of the selection of a distinct form of a seedling, hybrid or wild plant which is maintained by cultivation
Cuneate wedge-shaped (glossary fig. 4)
Deciduous bearing leaves which fall on a regular basis, usually annually
Decumbent growing horizontally but becoming erect at tips
Dehisced split open, usually referring to the anthers split to release the pollen, or the fruits to release the seeds
Dentate of a leaf margin with sharp outwardly pointing teeth (glossary fig. 12)
Dichotomous regularly forked; divided into two equal branches
Diploid a double set of chromosomes, as is usually found in the vegetative cells of a plant
Dormancy (adj. **dormant**) a period of inactivity to survive adverse conditions
Elliptic about twice as long as wide, tapering to each end (glossary fig. 24)
Endemic a species which is only found in one particular geographical area
Entire referring to margin of a leaf, bract or petal which is not lobed or toothed (glossary fig. 9)
Exserted sticking out, usually referring to the stamens of a flower
Filament the stalk bearing the anther (fig. 7)

Filiform thread-like
Fruit the structure containing the seeds
Genus (pl. **genera**) a botanical category used to group together related species
Geophyte a plant which has underground tubers for food storage
Glabrous without hairs
Glaucous bluish or greyish-green
Habit the appearance of the plant
Habitat the situation where a species may be found growing in the wild
Haploid a single set of chromosomes usually found in the pollen or ovules
Herbaceous a plant with non-woody stems which do not persist from year to year. Many pelargoniums have soft non-woody growth but do not necessarily die down to ground level each year; they are referred to here as herbaceous to indicate their fleshy rather than woody stems
Hispid with stiff short hairs usually making a leaf or stem feel rough to the touch
Hybrid the plant formed as a result of the deliberate or accidental crossing of two different species or cultivars. A bigeneric hybrid is created when two genera are crossed
Hypanthium in the genus *Pelargonium*, an extension of the upper sepal modified to form a tube, fused to the pedicel, which produces nectar attracting insects for pollination (fig. 7)
Inflorescence the flowering stem bearing one or more flowers grouped together (fig. 6)
Internode the space between two nodes (fig. 6)
Irregular not radially symmetrical, usually referring to flower (fig. 5)
Keel a ridge, often referring to the midrib on the underside of the leaf or of another organ
Lamina leaf blade (fig. 6)
Lanceolate lance-shaped, three times as long as wide tapering to tip (glossary fig. 20)
Lax arranged loosely
Leaflet one part of a compound leaf
Linear more than five times as long as wide with parallel sides (glossary fig. 19)
Mericarp a one seeded section of the fruit which separates to release the seeds (fig. 4)
Meristem the area of a plant where the cells are actively dividing. The apical meristem of a shoot may be dissected out and cultured to create new plants. This process is called meristem culture
Microclimate the environmental conditions in the specific area in which a plant lives
Morphology (adj. **morphological**) the study of the external features of a plant
Nectary spur see hypanthium (fig. 7)
Node the junction from which the leaves arise (fig. 6)
Oblanceolate lanceolate but tapering towards leaf stalk (glossary fig. 21)
Oblong up to five times as long as wide with parallel sides (glossary fig. 23)
Obovate ovate but tapering towards leaf stalk (glossary fig. 26)
Obtuse blunt (glossary fig. 3)
Orbicular circular
Ovary the female part of the flower containing the ovules which after fertilization develop into seeds (fig. 7)
Ovate egg-shaped, broadest at base (glossary fig. 25)
Palmate a leaf with three or more veins or leaflets arising from the apex of the leaf stalk (glossary figs. 16 & 18)

Palmatifid a leaf divided palmately from the margin about halfway to the midrib (glossary fig. 31)

Palmatisect a leaf divided palmately from the margin almost to the midrib (glossary fig. 32)

Pedicel the stalk of an individual flower (fig. 6)

Peduncle the stalk of an inflorescence (fig. 6)

Peltate refers to leaf when the petiole is attached to the centre of the lamina (glossary fig. 8)

Perennial a plant persisting for more than two years

Persistent used to refer to stipules or petioles which do not fall off as they die

Petiole leaf stalk (fig. 6)

Photosynthesis the process by which a plant is able to produce energy for growth

Pinna (pl. **pinnae**) individual leaflet of a pinnate leaf

Pinnate a leaf with the leaflets or veins arranged on either side of an axis (glossary figs. 15 & 17)

Pinnatifid a leaf divided pinnately from the margin about halfway to the midrib (glossary fig. 29)

Pinnatisect a leaf divided pinnately from the margin almost to the midrib (glossary fig. 30)

Posterior organ closest to stem, usually referring to upper sepal or petals (fig. 6)

Pubescent with short hairs

Rachis the main axis of a compound leaf

Regular symmetrical, usually referring to flower (fig. 5)

Reniform kidney-shaped (glossary fig. 27)

Reticulate netted, usually referring to the pattern of the veins

Revolute rolled under, usually referring to the leaf margin

Rosette a plant with a short stem from which the leaves arise in a spiral pattern and often lying horizontally to the ground

Scandent Climbing

Scape the stem bearing the inflorescence which arises directly from the ground

Section a subdivision within the genus

Sepal a segment of the calyx (fig. 7)

Serrate of a leaf margin with sharp forwardly pointing teeth (saw-like) (glossary fig. 13)

Sessile without a stalk

Simple referring to a leaf which is not divided into individual leaflets (glossary figs. 15 & 16) or to an unbranched inflorescence

Spathulate spoon-shaped (glossary fig. 22)

Species a group of individuals with the same basic characteristics which are found in the wild and breed true from seed

Stamen the male part of the flower-bearing anthers (fig. 7)

Staminode an infertile stamen (fig. 7)

Stigma the female part of the flower which receives pollen (fig. 7)

Stipule a small leaf-like appendage formed at the base of the leaf stalk (fig. 6)

Style the stalk extending from the ovary bearing the stigma (fig. 7)

Subshrub a plant with stems which are woody at base but with non-woody herbaceous stems above

Subtending used for an organ of a plant in whose axil another organ arises, as in the case of a bract subtending a flower stem

Taxon (pl. **taxa**) a single group in a system of classification, for example, a genus, species or family. All members of a taxon have similar characteristics

Terete more or less circular in cross section

Tomentose thickly covered with hairs

Transpiration the process by which the plant loses water through the leaves

Trifid divided into three but not cut completely to base (glossary fig. 33)

Trifoliate a compound leaf divided into three leaflets

Truncate cut across at right angles (glossary fig. 5)

Tuber a swollen underground organ used for food storage

Umbel an inflorescence where all the flower stalks arise from the top of the peduncle (fig. 6). The youngest flowers are found in the centre of the inflorescence. In pelargoniums the oldest flowers are found in the centre of the inflorescence and the inflorescence is sometimes referred to as a pseudoumbel

Viscous (adj. **viscid**) sticky

Whorl a group of three or more similar organs, for example, leaves, arising from the same level on the stem and arranged in a circle

Xerophyte (adj. **xerophytic**) a plant adapted to survive drought or very dry situations

Zygomorphic see under irregular (fig. 5)

ENDNOTES

1. VAN DER WALT, J.J.A. and ROUX, J.P., *Taxonomy and phylogeny of* Pelargonium *section* Campylia *(Geraniaceae)*, South African Journal of Botany, **57**: 1991, pp. 291–4.

2. McDONALD, D.J. and VAN DER WALT, J.J.A., *Observations on the pollination of* Pelargonium tricolor, *section* Campylia *(Geraniaceae)*, South African Journal of Botany, **58**: 1992, pp. 386–92.

3. VAN DER WALT, J.J.A. and VAN ZYL, L., *A taxonomic revision of* Pelargonium *section* Campylia *(Geraniaceae)*, South African Journal of Botany, **54**: 1988, pp. 145–71.

4. VAN DER WALT, J.J.A., VENTER, H.J.T., VERHOEVEN, R. and DREYER, L.L., Erodium incarnatum *transferred to the genus* Pelargonium *(Geraniaceae)*, South African Journal of Botany, **56**: 1990, pp. 560–4.

5. GIBBY, M., 'Pelargonium ranunculophyllum *(Geraniaceae) in southern Africa*', South African Journal of Botany, **55**: 1989, pp. 539–42.

6. OLIVIER, M.C., and VAN DER WALT, J.J.A., *The taxonomy of the* Pelargonium peltatum *(L.) L'Hérit. complex*, Journal of South African Botany, **50**: 1984, pp. 1–14.

7. BROWN, N.E., *New or noteworthy plants*, Gardeners Chronicle, **8**: 1890, pp. 154.

8. SWEET, R., Hortus Brittanicus, (1839) p. 120.

9. BOLUS, L., *Plants – New and Noteworthy* South African Gardening and Country Life, **XVII**: 1927, pp. 241.

10. DYER, R.A., Pelargonium x kewense, Journal of Royal Horticultural Society, **LIX**: 1934, pp. 212–15.

11. DREYER, L.L., ALBERS, F., VAN DER WALT, J.J.A. and MARSCHEWSKI, D.E., *Subdivision of* Pelargonium *section* Cortusina *(Geraniaceae)*, Plant Systematics and Evolution, **183**: 1992, pp. 83-97.

12. VAN DER WALT, J.J.A., ALBERS, F., GIBBY, M., *Delimitation of* Pelargonium *sect.* Glaucophyllum *(Geraniaceae)*, Plant Systematics and Evolution, **171**: 1990, pp.15–26.

13. VAN DER WALT, J.J.A., The flowering plants of Africa, 1991, plate 2035.

14. JACQUIN, N.J., Icones Plantarum Rariorum, **3**: 1792, plate 539 *P. cortusaefolium*; plate 540 *P. fuscatum*; plate 541 *P. patulum*.

[15] VAN DER WALT, J.J.A., Pelargoniums of Southern Africa, **2**: 1981, pp. 87.

[16] COETZEE, N., and VAN DER WALT, J.J.A., *Three varieties of* Pelargonium patulum *(Geraniaceae)*, Journal of South African Botany, **58**: 1992, pp. 77-82.

[17] MARAIS, E.M., *A taxonomic revision of the* Pelargonium pinnatum *complex (Geraniaceae)*, South African Journal of Botany, **59**: 1993, pp. 123–134.

[18] SCHELTEMA, A.G., and VAN DER WALT, J.J.A., *Taxonomic revision of* Pelargonium *section* Jenkinsonia *(Geraniaceae) in southern Africa*, South African Journal of Botany, **56**: 1990, pp. 285–301.

[19] ALBERS, F., GIBBY, M., and AUST-MANN, M., *A reappraisal of* Pelargonium *sect.* Ligularia *(Geraniaceae)*, Plant Systematics and Evolution, **179**: 1992, pp. 257-276.

[20] VAN DER WALT, J.J.A. and BOUCHER, D.A., *A taxonomic revision of the section* Myrrhidium *of* Pelargonium *(Geraniaceae) in southern Africa*, South African Journal of Botany **52**: 1986, pp. 438–62.

[21] VAN DER WALT, J.J.A., *A taxonomic revision of the section of* Pelargonium *L'Hérit. (Geraniaceae)*, Bothalia, **15**: 1985, pp. 345–385.

[22] VOLSCHENK, B., VAN DER WALT, J.J.A. and Vorster, P.J., *The subspecies of* Pelargonium cucullatum *(Geraniaceae)*, Bothalia **14**: 1982, pp. 45-51.

[23] CAROLIN, R.C., *The genus* Pelargonium *L'Héritier ex Ait. in Australia*, Proceedings of the Linnean Society of New South Wales, **86**: 1961 pp. 280–294.

[24] VAN DER WALT, J.J.A., and VORSTER, P.J., *Name changes in* Pelargonium *(Geraniaceae)*, Journal of South African Botany, **46**: 1980, pp. 283–292.

SELECT BIBLIOGRAPHY

Abbott, P., *The Peter Abbott guide to the scented Geraniaceae*, Hill, 1994.

Clark, D., *Pelargoniums*, Collingridge, 1988.

Clifford, D., *Pelargoniums including the popular geranium*, Blandford Press, 1970.

Gunn, M. & CODD, L.E., *Botanical exploration of southern Africa*, A.A. Balkema, 1981.

Taylor, J., *Geraniums and pelargoniums*, Crowood Press, 1989.

Van der Walt, J.J.A., *Pelargoniums of southern Africa*, Vol. 1, Purnell, 1977.

Van de Walt, J.J.A. & Vorster, P.J., *Pelargoniums of southern Africa*, Vol. 2, National Botanic Gardens, Kirstenbosch, 1981.

Van de Walt, J.J.A. & Vorster, P.J., *Pelargoniums of southern Africa*, Vol. 3, Juta, 1988.

Webb, W.J., *The Pelargonium family*, Croom Helm, 1984.

Other historical references and floras mentioned in the text should be available from specialist libraries. Journals that may also be of relevance have been included. Specific references to articles mentioned in endnotes, references to the first publications of names and references to illustrations have not been detailed here. These may be found under the species descriptions.

Aiton, W., *Hortus Kewensis*, London, 1789.

Andrews, H.C., *The botanist's repository*, London, 1797–1814.

Andrews, H.C., *Geraniums*, 2 Volumes, London, 1805-1806.

Bothalia, Pretoria, 1921 to present day.

Burman, J., *Rariorum africanarum plantarum*, Amsterdam 1738.

Carolin, R.C., *The genus* Pelargonium *L'Hér. ex Ait. in Australia*, Proceedings of the Linnean society of New South Wales, **86**: 1962, pp. 280–294.

Cavanilles, A.J., *Monadelphiae classis dissertationes*, **4**: Paris, 1787.

Commelin, C., *Hortus medicus Amstelodamensis*, Praelydia botanica, Amsterdam, 1689.

Commelin, C., *Icones plantarum horti medici Amstelaedamensis*, Amsterdam, 1716.

Curtis, W., *Botanical magazine* (now *Kew magazine*), London 1787, to date.

De Candolle, A.P., *Prodromus systematis naturalis regni vegetabilis*, **1**: Paris, 1824.

Dillenius, J.J., *Hortus Elthamensis*, London, 1732.

Don, G., *A general system of gardening and botany*, **1**: London, 1831.

Edwards, S., *The botanical register*, London, 1815-1847.

Ehrhart, F., *Beiträge zur Naturkunde*, Hanover, 1787.

Flowering plants of South Africa, Pretoria, 1921–1944; (from 1945 as The flowering plants of Africa).

Harvey, W.H., *Flora Capensis*, **1**: Dublin, 1860.

Hermann, P., *Horti academica Lugdunobatavi catalogus*, Leiden, 1687.

Hermann, P., *Paradisi batavus*, Amsterdam, 1705.

Index Kewensis, Oxford, 1895 to date.

Jacquin, N.J., *Icones plantarum rariorum*, Vienna, 1781-1795.

Jacquin, N.J., *Plantarum rariorum horti caesarei Schoenbrunnensis*, Vienna, 1797-1804.

Journal of South African botany, South Africa, 1935–1984; (from 1985 as *South African journal of botany*).

Knuth, R., *Das Pflanzenreich* 4, **129:** Berlin, 1912.

Kokwaro, J.O., in *Flora of tropical East Africa*, Geraniaceae, Crown Agents, London, 1971.

L'Hértier, C. L., *Geraniologia*, Paris, 1792.

Linnaeus, C., *Species plantarum*, Stockholm, 1753.

Loddiges, C., *The botanical cabinet*, London, 1818–1833.

Moore, H.E., *Pelargoniums in cultivation*, Baileya, **3:** New York, 1955.

Müller, T., in *Flora Zambesiaca*, **2:** Crown Agents, London, 1963.

Steudel, E.T., *Nomenclator botanicus*, Stuttgart, 1841.

Sweet, R., *Geraniaceae*, **1–5:** London, 1820–1830.

Sweet, R., *Hortus Brittanicus*, London, 1827.

Willdenow, C.L., *Species plantarum*, Berlin, 1800.

Willdenow, C.L., *Hortus Berolinensis*, Berlin, 1806-1816.

Willdenow, C.L., *Enumeratio plantarum*, Berlin, 1809.

■ SPECIES INDEX

A list of species arranged according to their botanical sections. Synonyms are not included here.

Photographic plate numbers (not page numbers) are indicated by italics. Line drawing numbers (not page numbers) are indicated by bold italics.

Campylia (Sweet) de Candolle 35, **36–41**
caespitosum Turcz. **37**
capillare (Cav.) Willd. 36, **37–38**, 39
coronopifolium Jacq. **38**, 39
elegans (Andr.) Willd. 37, **38–39**, 40
incarnatum (L'Hérit.) Moench 36, **39**, 40
ocellatum J.J.A. van der Walt **36**, 39
oenothera (L.f.) Jacq. **38**, 39
ovale (Burm. f.) L'Hérit. *1, 3,* 36, 38, **40**
setulosum Turcz. 39, **40**, 75
tricolor Curt. *2, 3,* 36, 38, 39, 40, **141**

Ciconium (Sweet) Harvey 28, 35, **42–55**, 75, 77, 81
acetosum (L.) L'Hérit. 26, **43**, 51
acreaum R.A. Dyer **44**
alchemilloides (L.) L'Hérit. 42, **44**, 49
caylae Humbert *4,* 42, **45**
elongatum (Cav.) Salisb. 26, 42, **45–46**, 64
frutetorum R.A. Dyer *5,* **46**, 51
hararense Engl. **48**
inquinans (L.) L'Hérit. *6,* **9**, 42, **46–48**, 51, 55, 135
multibracteatum Hochst. 46, **48**, 53
mutans Vorster **48–49**, 61, 135
peltatum (L.) L'Hérit. *7,* 26, 42, **49–50**, 52, 136, 149
quinquelobatum Hochst. *8,* 50–51
ranunculophyllum (Eckl. & Zeyh.) Bak. **44**
salmoneum R.A. Dyer *9,* 46, **51**, 52, 135

stenopetalum Ehrh. *9,* **52**
tongaense Vorster **52–53**

transvaalense Knuth **53**
usambarense Engl. **48**
zonale (L.) L'Hérit. *10,* **10**, 17, 26, 42, 46, 51, **53–55**

Cortusina (de Candolle) Harvey 12, 28, 35, **55–60**, 75, 115, 125, 130
cortusifolium L'Hérit. 18, **57**, 59, 115, 130
crassicaule L'Hérit. 18, **57–58**, 132
desertorum Vorster 56, **58**, 60
echinatum Curt. *11,* 55, 56, **58–59**, 130
magenteum J.J.A. van der Walt **59**, 84, 130
mirabile Dinter **58**
sibthorpiifolium Harv. 18, **59–60**
xerophyton Schltr. ex Knuth *12,* 56, 58, 60

Glaucophyllum Harvey 35, 56, **60–64**, 79, 95, 96, 110
fruticosum (Cav.) Willd. **61**, 62, 64
grandiflorum (Andr.) Willd. *13,* **61–62**, 63
laevigatum (L.f.) Willd. *14,* 61, **62–63**, 64
lanceolatum (Cav.) Kern *15,* **63**, 110
patulum Jacq. 39, 46, 60 ,62, **63–64**
tabulare (Burm. f.) auct. non L'Hérit. *16,* 46, 62, 63, **64**
ternatum (L.f.) Jacq. 61, 62, **64**

Hoarea (Sweet) de Candolle 28, 35, **64–70**, 77, 118
appendiculatum (L.f.) Willd. **66**
asarifolium (Sweet) G. Don **66–67**
auritum (L.) Willd. **67**
incrassatum (Andr.) Sims 66, **67–68**
longifolium (Burm. f.) Jacq. *17,* **68**

oblongatum E. Mey. ex Harv. **68**, 69
pinnatum (L.) L'Hérit. *18*, 65, **68–69**, 139
punctatum (Andr.) Willd. **69**
rapaceum (L.) L'Hérit. *19*, 66, **69–70**
trifoliolatum (Eckl. & Zeyh.) E.M. Marais **69**
viciifolium DC **69**

Isopetalum (Sweet) de Candolle 35, **70**
cotyledonis (L.) L'Hérit. *20*, **70**, 91

Jenkinsonia (Sweet) Harvey 28, 35, **70–74**, 87
antidysentericum (Eckl. & Zeyh.) Kostel. **71–72**, 73
boranense Friis & Gilbert 71, **72**
endlicherianum Fenzl *21*, 71, **72–73**, 133
praemorsum (Andr.) Dietr. 28, 71, 72, **73**
quercetorum Agnew *22*, 71, 72, **73–74**, 87
tetragonum (L.f.) L'Hérit. *22*, 71, **74**, 87

Ligularia Harvey 28, 35, 56, **74–86**, 125
alpinum Eckl. & Zeyh. *24*, 75, **76**
aridum R.A. Dyer 42, 75, **76–77**
articulatum (Cav.) Willd. 42, 75, **77**
barklyi Scott Elliot 42, 75, **77**
crassipes Harv. 75, **78**, 81, 82
divisifolium Vorster *25*, 75, **78**
dolomiticum Knuth 75, **78–79**, 80, 85
exhibens Vorster 75, **79**, 86
fulgidum (L.) L'Hérit. 75, **79–80**, 122
grandicalcaratum Knuth 75, **80**, 82
griseum Knuth 75, 79, **80**, 85
hirtum (Burm. f.) Jacq. 75, **80**
hystrix Harv. 75, 78, **81**, 82
karooicum Compton & Barnes 75, **81**
mollicomum Fourc. 75, **81**
oreophilum Schltr. 75, 78, 80, **82**
otaviense Knuth 75, 80, **82**, 84
plurisectum Salter 75, **82–83**, 126
pulchellum Sims 75, **83**, 85
rodneyanum Mitchell ex Lindl. *26*, **83–84**

sericifolium J.J.A. van der Walt 75, **84**
spinosum Willd. *27*, 75, 82, **84**
stipulaceum (L.f.) Willd. *28*, 75, **84–85**
tenuicaule Knuth 75, **85**
tragacanthoides Burch. 75, 80, **85–86**
trifidum Jacq. *29*, 75, 79, **86**
worcesterae Knuth 75, 81, **86**

Myrrhidium de Candolle 35, 71, **87–90**
candicans Spreng. **88**
caucalifolium Jacq. **88**
gallense Chiov. **89**
goetzeanum Engl. **89**
longicaule Jacq. *30*, 87, **88–89**, 90
multicaule Jacq. **89**
myrrhifolium (L.) L'Hérit. 28, 70, 87, 88, **89–90**
suburbanum Clifford ex Boucher *31*, **90**
whytei Baker **89**

Otidia (Sweet) G. Don 28, 35, 70, **90–95**
alternans Wendl. 91, **92**, 94
carnosum (L.) L'Hérit. *27*, 91, **92**
ceratophyllum L'Hérit. *18*, **92–93**
crithmifolium J.E. Sm. 91, **93**
dasyphyllum E. Mey. ex Knuth *32*, **93–94**
klinghardtense Knuth *18*, **94**
laxum (Sweet) G. Don **94**
paniculatum Jacq. **94–95**

Pelargonium de Candolle 12, 13, 35, 56, 60, 79, **95–112**
betulinum (L.) L'Hérit. 13, *33*, 37, 96, 97, 100
capitatum (L.) L'Hérit. 95, 96, **97–98**, 104, 109, 112, 113, 114
citronellum J.J.A. van der Walt *34*, **98**, 110
cordifolium (Cav.) Curt. 35, **98–99**
crispum (Berg.) L'Hérit. **99**, 106, 116, 136
cucullatum (L.) L'Hérit. 12, 16, *36*, 37, 62, 95, 97, **99–100**, 135
denticulatum Jacq. **100–101**, 102, 108, 109

englerianum Knuth 99, **101**, 110, 111
glutinosum (Jacq.) L'Hérit. **101–102**
'Graveolens' *11*, *38*, **103–104**, 109
graveolens auct. non L'Hérit. *38*, 95, **102**
greytonense J.J.A. van der Walt **104–105**
hermanniifolium (Berg.) Jacq. *39*, 99, 104, **106**
hispidum (L.f.) Willd. **106**, 107
panduriforme Eckl. & Zeyh. *40*, **106–107**, 108, 135
papilionaceum (L.) L'Hérit. 104, 106, **107**
pseudoglutinosum Knuth *40*, **107–108**
quercifolium (L.f.) L'Hérit. *40*, 107, **108**
radens H.E. Moore *38*, 95, 96, 100, 103, 104, **108–109**
ribifolium Jacq. **109–110**
scabroide Knuth **110**
scabrum (Burm. f.) L'Hérit. 98, **110**, 144
sublignosum Knuth *41*, **110–111**
tomentosum Jacq. 41, **111**, 136
vitifolium (L.) L'Hérit. 95, 97, 98, 107, **111–112**

Peristera de Candolle 35, **112–118**
acugnaticum Thou. **116**
apetalum P. Taylor **116**
australe Willd. 18, *42*, **113–114**, 115, 116, 128, 139
australe Willd. 'Redondo' **114**
brevirostre R.A. Dyer **117**
chamaedryfolium Jacq. **114**
columbinum Jacq. 112, 113, **114–115**
drummondii Turcz. 113, 114, **115**
glechomoides A. Rich. **116**
grossularioides (L.) L'Hérit. **115–116**
helmsii Carolin **116**
hypoleucum Turcz. **116**
inodorum Willd. **116**
iocastum (Eckl. & Zeyh.) Steud. *43*, **116–117**
littorale Hügel **116**
madagascariense Baker **116**

minimum (Cav.) Willd. 112, **117**
nanum L'Hérit. **117**
nelsonii Burtt Davy **117**
pseudofumarioides Knuth **117**
renifolium Swinbourne **116**
senecioides L'Hérit. **117–118**

Polyactium de Candolle 28, 35, 65, 77, 79, 83, **118–125**
anethifolium (Eckl. & Zeyh.) Steud. **119**
bowkeri Harv. *44*, **119–120**, 121
caffrum (Eckl. & Zeyh.) Harv. *45*, **120–121**
gibbosum (L.) L'Hérit. *46*, 118, **121**, 123
lobatum (Burm. f.) L'Hérit. **121–122**, 139
luridum (Andr.) Sweet *47*, 118, **122–123**
multiradiatum Wendl. *48*, 118, **123**
pulverulentum Colv. ex Sweet 118, **123**, 124
radulifolium (Eckl. & Zeyh.) Steud. 122, **123–124**
schizopetalum Sweet *49*, **124**
triste (L.) L'Hérit. 15, 25, 66, 118, 119, 122, 123, **124–125**

Reniformia (Knuth) Dreyer 12, 28, 35, 55, 56, 75, 82, **125–130**
abrotanifolium (L.f.) Jacq. 75, 78, 82, **126**, 127
album J.J.A. van der Walt **126–127**
dichondrifolium DC. 86, 125, **127**, 128, 129
exstipulatum (Cav.) L'Hérit. *50*, 126, **127**, 128
fragrans Willd. *51*, **128**
ionidiflorum (Eckl. & Zeyh.) Steud. *52*, 125, **128**
mossambicense Engl. **126–127**
odoratissimum (L.) L'Hérit. *51*, 81, 125, 126, 127, 128, **129**, 136
reniforme auct. non Curtis *53*, 59, **129–130**
sidoides DC. *54*, 129, **130**

■ PLANT INDEX

A total alphabetical list of plants. The majority of synonyms likely to be found are included here. Most of the older synonyms using the generic name *Geranium* have been omitted as usually there has been a direct substitution with the correct generic name, *Pelargonium*. Likewise, direct substitutions of *Pelargonium* for an earlier generic name now used for the section name, have also been excluded.

Campylia blattaria (Jacq.) Sweet = *P. ovale* (Burm. f.) L'Hérit. 40
Campylia veronicifolia Eckl. & Zeyh = *P. ovale* (Burm. f.) L'Hérit. subsp. *veronicifolium* 40
Chorisma Eckl. & Zeyh. see section **Jenkinsonia** 71
Chorisma flavescens Eckl. & Zeyh. = *P. tetragonum* (L.f.) L'Hérit. 71

Dimacria Sweet see section **Hoarea** 64

Erodium 22, 23, 26, 39, 111
Erodium hymenodes L'Hérit. = *Erodium trifolium* (Cav.) Cav. 114
Erodium incarnatum (L.) L'Hérit. = *P. incarnatum* (L'Hérit.) Moench 39
Erodium pelargoniiflorum Boiss. & Heldr. 114
Erodium trifolium (Cav.) Cav. 114

Geraniaceae 21, 27
Geraniospermum Knutze see section **Myrrhidium** 87
Geranium 22, 26

Geranium odoratissimum erectum Andr. = *P. fragrans* Willd. 128
Geranium pictum Andr. = *pulchellum* Sims 83
Geranium pinnatum Cav. = *P. viciifolium* DC 69
Geranium selinum Andr. = *rapaceum* (L.) L'Hérit. 69
Geranium villosum Andr. = *P. tomentosum* Jacq. 111
Geranium viscosum Cav. = *P. glutinosum* (Jacq.) L'Hérit. 101
Grenvillea Sweet see section **Hoarea** 64

Hoarea corydalifolium Sweet = *rapaceum* (L.) L'Hérit. 69

Jenkinsonia see section **Jenkinsonia** or section **Myrrhidium** 71, 87
Jenkinsonia pendula Sweet = *P. longicaule* Jacq. var. *longicaule* 88
Jenkinsonia quinata (Sims) Sweet = *P. praemorsum* (Andr.) Dietr. 73
Jenkinsonia synnoti Sweet = *P. myrrhifolium* (L'Hérit.) var. *myrrhifolium* 89

Monsonia 21
Myrrhidium urbanum Eckl. & Zeyh. = *P. suburbanum* Clifford ex Boucher 90

Otidia dasycaulon Eckl. & Zeyh. = *P. crithmifolium* J.E. Sm. 93
Otidia ferulacea DC. ex Eckl. & Zeyh. = *P. carnosum* (L.) L'Hérit. 92

Phymatanthus see section **Campylia** 36
Phymatanthus elatus Sweet = *P. tricolor* Curt. 41

Sarcocaulon 22
Seymouria Sweet see section **Hoarea** 64

Pelargonium
abrotanifolium (L.f.) Jacq. 75, 78, 82, **126**, 127
acerifolium L'Hérit. = *cucullatum* (L.) L'Hérit. subsp. *strigifolium* Volschenk 100
acetosum (L.) L'Hérit. 26, **43**
aconitifolium (Eckl. & Zeyh.) Steud. = *luridum* (Andr.) Sweet 122
acreaum R.A. Dyer 44
acugnaticum Thou. see under *grossularioides* (L.) L'Hérit. 116
album J.J.A. van der Walt 126
alchemillifolium Salis. = *alchemilloides* (L.) L'Hérit. 44
alchemilloides (L.) L'Hérit. 42, **44**, 49
alpinum Eckl. & Zeyh. 24, 75, **76**
alternans Wendl. 91, **92**, 94
amabile Dinter = *sibthorpiifolium* Harv. 59
amatymbicum (Eckl. & Zeyh.) Harv = *schizopetalum* Sweet 124
anceps Soland. = *grossularioides* (L.) L'Hérit. 115
anethifolium (Eckl. & Zeyh.) Steud. **119**
angulosum (Mill.) L'Hérit. = *cucullatum* (L.) L'Hérit. subsp. *cucullatum* 100
angustissimum E. Mey. ex Knuth = *coronopifolium* Jacq. 38
antidysentericum (Eckl. & Zeyh.) Kostel. subsp. *antidysentericum* **71–72**, 73
antidysentericum (Eckl. & Zeyh.) Kostel. subsp. *inerme* Scheltema **72**
antidysentericum (Eckl. & Zeyh.) Kostel. subsp. *zonale* Scheltema **72**

apetalum P. Taylor see under *grossularioides* (L.) L'Hérit. **116**
appendiculatum (L.f.) Willd. **66**
aridum R.A. Dyer 42, 75, **76–77**
armatum Sweet = a pink-flowered form of *echinatum* Curt. 58
artemisifolium sensu auct. non DC = *divisfolium* Vorster 78
articulatum (Cav.) Willd. 42, 75, **77**
asarifolium (Sweet) G. Don **66–67**
asperum Willd. = 'Graveolens' 103
astragalifolium (Andr.) Loudon = *trifoliolatum* (Eckl. & Zeyh.) E.M. Marais 69
augustissimum E. Mey ex Knuth = *coronopifolium* Jacq. 38
auritum (L.) Willd. subsp *auritum* **67**
auritum (L.) Willd. subsp. *carneum* (Harv.) J.J.A. van der Walt **67**
australe Willd. 18, *42*, **113–114**, 115, 116, 128, 139
australe Willd. 'Redondo' **114**

bachmannii Knuth = 'Saxifragoides' 52
balsameum Jacq. = *scabrum* (L.) L'Hérit. 110
barbatum Jacq. = *longifolium* (Burm. f.) Jacq. 68
barklyi Scott Elliot 42, 75, **77**
bechuanicum Burtt Davy = *dolomiticum* Knuth 79
betonicum (Burm. f.) Jacq. = *myrrhifolium* (L.) Herit. var. *myrrhifolium* 89
betulinum (L.) L'Hérit.13, *33*, 37, **96–97**, 100
blattarium Jacq. = *ovale* (Burm. f.) L'Hérit. 40
boranense Friis & Gilbert 71, **72**
bowkeri Harv. 44, **119–120**, 121
brevipetalum N.E. Brown = parviflorum Wendl. 92

brevirostre R.A. Dyer see under *nanum* L'Hérit. **117**
bullatum Jacq. = *myrrhifolium* (L.) Herit. var. *myrrhifolium* 89
burtoniae L. Bolus = *stenopetalum* Ehrh. 52
caespitosum Turcz. subsp. *caespitosum* **37**
caespitosum Turcz. subsp. *concavum* Hugo **37**
caffrum (Eckl. & Zeyh.) Harv. 45, **120–121**
canariense Willd. = *candicans* Spreng. 88
candicans Spreng. **88**
capillare (Cav.) Willd. 36, **37–38**, 39
capitatum (L.) L'Hérit. 95, 96, **97–98**, 104, 109, 112, 113, 114
capnoides L'Hérit. = *minimum* (Cav.) Willd. 114
cardiophyllum Harv. = *setulosum* Turcz. 40
carnosum (L.) L'Hérit. 27, 91, **92**
caucalifolium Jacq. subsp. *caucalifolium* **88**
caucalifolium Jacq. subsp. *convolvulifolium* (Schltr. ex Knuth) J.J.A. van der Walt **88**
caylae Humbert 4, 42, **45**
ceratophyllum L'Hérit. 18, **92–93**
chamaedryfolium Jacq. **114**
ciliatum (Cav.) Jacq. = *longifolium* (Burm.f.) Jacq. 68
citronellum J.J.A. van der Walt 34, **98**, 110
clandestinum L'Hérit. = *inodorumum* Willd. 116
clypeatum (Eckl. & Zeyh.) Steud. = *peltatum* (L.) L'Hérit. 50
columbinum Jacq. 112, 113, **114–115**
convolvulifolium (Schltr.) ex Knuth = *caucalifolium* Jacq. subsp. *convolvulifolium* (Schltr. ex Knuth) J.J.A. van der Walt 88
cordatum L'Hérit. = *cordifolium* (Cav.) Curt. 98
cordifolium (Cav.) Curt. 35, **98–99**
cordifolium (Cav.) Curt. var. *lanatum* Harv. **99**

cordifolium (Cav.) Curt. var. *rubrocinctum* Harv. **99**
coriandrifolium Jacq. = *myrrhifolium* (L.) L'Hérit var. *coriandrifolium* (L.) Harv. 90
coronopifolium Jacq. **38**, 39
coronopifolium Jacq. var. lineare Harvey = *coronopifolium* Jacq. 38
cortusaefolium Jacq. see under *grandiflorum* (Andr.) Willd. 62
cortusifolium L'Hérit. 18, 57, **59**, 115, 130
cotyledonis (L.) L'Hérit. 20, **70**, 91
craddockense Knuth = *dichondrifolium* DC. 127
crassicaule L'Hérit. 18, **57–58**, 132
crassipes Harv. 75, **78**, 81, 82
crispum (Berg.) L'Hérit. **99**, 106, 116, 136
crispum (Berg.) L'Hérit. 'Major' **99**
crispum (Berg.) L'Hérit. 'Minor' **99**
crispum (Berg.) L'Hérit. 'Variegatum' **99**
crithmifolium J.E. Sm. 91, **93**
cucullatum (L.) L'Hérit. subsp. *cucullatum* 12, 16, 36, 62, 95, 97, **99–100**, 135
cucullatum (L.) L'Hérit. subsp. *strigifolium* Volschenk **99**
cucullatum (L.) L'Hérit. subsp. *tabulare* Volschenk 37, **99**
cucullatum (L.) L'Hérit. 'Flore Plenum' **100**

dasycaulon Sims = *crithmifolium* J.E. Sm. 93
dasyphyllum E. Mey. ex Knuth 32, **93–94**
daucifolium Steud. = *triste* (L.) L'Hérit. 124
denticulatum Jacq. **100–101**, 102, 108, 109
desertorum Vorster 56, **58**, 60
dichondrifolium DC. 86, 125, **127**, 128, 129
dissectum (Eckl. & Zeyh.) Harvey = *aridum* R.A. Dyer 76
divaricatum (Thunb.) DC. = *fruticosum* (Cav.) Willd. 61

diversifolium Wendl. see under *laevigatum* (L.f.) Willd. 63
divisifolium Vorster 25, 75, **78**
dolomiticum Knuth 75, **78–79**, 80, 85
drummondii Turcz. 113, 114, **115**

echinatum Curt. *11*, 55, 56, **58–59**, 130
echinatum Curt. 'Album' = *echinatum* Curt. 58
elegans (Andr.) Willd. 37, **38–39**, 40
elongatum (Cav.) Salisb. 26, 42, **45–46**, 64
endlicherianum Fenzl *21*, 71, **72–73**, 133
englerianum Knuth 99, **101**, 110, 111
eriostemon Jacq. = *ovale* (Burm. f.) L'Hérit. 40
erodioides Hook = *australe* Willd. 113
exhibens Vorster 75, **79**, 86
exstipulatum (Cav.) L'Hérit. 50, 126, **127**, 128

ferulaceum var. *polycephalum* Harv. = *polycephalum* E. Mey. 92
ferulaceum (Burm. f.). Willd. = *polycephalum* E. Mey. 92
filipendulifolium Sweet = *triste* (L.) L'Hérit. 124
fischeri Engl. = *quinquelobatum* Hochst. 51
flabellifolium Harv. = *luridum* (Andr.) Sweet 122
flavum (Burm. f.) L'Hérit. = *triste* (L.) L'Hérit. 124
fragile (Andr.) Willd. = *trifidum* Jacq. 86
fragrans Willd. 51, **128**
fragrans Willd. 'Variegatum' **128**
frutetorum R.A. Dyer 5, **46**, 51
fruticosum (Cav.) Willd. **61**, 62, 64
fulgidum (L.) L'Hérit. 75, **79–80**, 122
fumarioides L'Hérit. = *minimum* (Cav.) Willd. 117
fuscatum Jacq. see under *grandiflorum* (Andr.) Willd. 62, 64

gallense Chiov. see under *multicaule* Jacq. **89**
gibbosum (L.) L'Hérit. 46, 118, **121**, 123
glaucum (L.f.) L'Hérit. = *lanceolatum* (Cav.) Kern 63
glechomoides A. Rich. see under *grossularioides* (L.) L'Hérit. 116
glomeratum (Andr.) Jacq. = *australe* Willd. 114
glutinosum (Jacq.) L'Hérit. **101–102**
goetzeanum Engl. see under *multicaule* Jacq. **89**
gramineum Bolus = *coronopifolium* Jacq. 38
grandicalcaratum Knuth 75, **80**, 82
grandiflorum (Andr.) Willd. *13*, **61–62**, 63
graveolens auct. non L'Hérit. 38, 95, **102**
greytonense J.J.A. van der Walt **104–105**
griseum Knuth 75, 79, **80**, 85
grossularioides (L.) L'Hérit. **115–116**

hararense Engl. see under *multibracteatum* Hochst. **48**
harveyanum Knuth = *hypoleucum* Turcz. 116
heckmannianum Engl. = *luridum* (Andr.) Sweet 122
hederinum Andr. = 'Lateripes' 50
helmsii Carolin see under *inodorum* Willd. **116**
heracleifolium Lodd. see under *lobatum* (Burm. f.) L'Hérit. 122, 124
hermanniifolium (Berg.) Jacq. *39*, 99, 104, **106**
heterogamum L'Hérit. see under *acraeum* 44
hirsutum (Burm. f.) Ait. var. *melananthum* (Jacq.) Harv. = *auritum* (L.) Willd. subsp. *auritum* 67
hirtum (Burm. f.) Jacq. 75, **80**
hispidum (L.f.) Willd. **106**, 107

hollandii Leighton = *pulverulentum* Colv. ex Sweet 123
humansdorpense Knuth 91
humifusum Willd. = *nanum* L'Hérit. 117
huraefolium Colv. = *luridum* (Andr.) Sweet 122
hybridum (L.) L'Hérit. see under *salmoneum* 51
hypoleucum Turcz. see under *grossularioides* (L.) L'Hérit. 116
hystrix Harv. 75, 78, **81**, 82
inaequilobum Mast. = *quinquelobatum* Hochst. 51
incarnatum (L'Hérit.) Moench 36, **39**, 40
incisum (Andr.) Willd. = *abrotanifolium* (L.f.) Jacq. 126
incrassatum (Andr.) Sims 66, **67–68**
inodorum Willd. 116
inquinans (L.) L'Hérit.
iocastum (Eckl. & Zeyh.) Steud. **9**, 6, 42, **46–48**, 51, 55, 135
ionidiflorum (Eckl. & Zeyh.) Steud. 52, 125, **128**

jacobii R.A. Dyer = *klinghardtense* Knuth 94

karooicum Compton & Barnes 75, **81**
klinghardtense Knuth 18, **94**

lacerum Jacq. = *myrrhifolium* (L.) Herit. var. *myrrhifolium* 89
laevigatum (L.f.) Willd. 14, 61, **62–63**, 64
lanceolatum (Cav.) Kern 15, **63**, 110
lateripes L'Hérit. = 'Lateripes' 50
laxum (Sweet) G. Don **94**
leptopetalum Sweet see under *stenopetalum* 52
littorale Hügel see under *inodorum* Willd. **116**
lobatum (Burm. f.) L'Hérit. **121–122**, 139

longicaule Jacq. var. *angustipetalum* Boucher **89**
longicaule Jacq. var. *longicaule* 30, 87, **88–89**, 90
longifolium (Burm. f.) Jacq. 17, **68**
luridum (Andr.) Sweet 47, 118, **122–123**

madagascariense Baker see under *grossularioides* (L.) L'Hérit. **116**
magenteum J.J.A. van der Walt **59**, 84, 130
malvaefolium Jacq. = *alchemilloides* (L.) L'Hérit. 44
melananthon Jacq. = *auritum* (L.) Willd. subsp. *auritum* 67
millefoliatum Sweet = *triste* (L.) L'Hérit. 124
minimum (Cav.) Willd. 112, **117**
mirabile Dinter see under *crassicaule* L'Hérit. **58**
mollicomum Fourc. 75, **81**
mossambicense Engl. see under *album* J.J.A. van der Walt **126–127**
multibracteatum Hochst. 46, **48**, 53
multicaule Jacq. subsp. *multicaule* **89**
multicaule Jacq. subsp. *subherbaceum* (Knuth) J.J.A. van der Walt **89**
multifidum Harv. = *plurisectum* Salter 82
multiradiatum Wendl. 48, 118, **123**
mutabile Sweet see under *P. gibbosum* (L.) L'Hérit. 121
mutans Vorster **48–49**, 61, 135
myrrhifolium (L.) Hérit. var. *coriandrifolium* (L.) Harv. 89, **90**
myrrhifolium var. longicaule (Jacq.) Harv. = *longicaule* Jacq. var. *longicaule* 88
myrrhifolium var. *myrrhifolium* (L.) L'Herit. 28, 70, 87, 88, **89–90**

nanum L'Hérit. **117**
nelsonii Burtt Davy see under *minimum* (Cav.) Willd. **117**

oblongatum E. Mey. ex Harv. **68**, 69
ocellatum J.J.A. van der Walt 36, **39**
odoratissimum (L.) L'Hérit. *51*, 81, 125, 126, 127, 128, **129**
oenothera (L.f.) Jacq. 38, **39**
oreophilum Schltr. 75, 78, 80, **82**
otaviense Knuth 75, 80, **82**, 84
ovale (Burm. f.) L'Hérit. subsp. *hyalinum* Hugo **40**
ovale (Burm. f.) L'Hérit. var. ovatum Harv. = *elegans* (Andrews) Willd. 38
ovale (Burm. f.) L'Hérit. subsp. *ovale* 1, 3, 36, 38, **40**
ovale (Burm. f.) L'Hérit. subsp. *veronicifolium* (Eckl. & Zeyh.) Hugo **40**
ovatostipulatum Knuth = *stipulaceum* (L.f.) Willd. 84
oxyphyllum DC. see under *laevigatum* (L.f.) Willd. 63

panduriforme Eckl. & Zeyh. *40*, **106–107**, 108, 135
paniculatum Jacq. **94–95**
papilionaceum (L.) L'Hérit. 104, 105, **107**
paradoxum Dinter = *klinghardtense* Knuth 94
parviflorum Wendl. = a form of *carnosum* (L.) L'Hérit. 92
parviflorum (Andr.) Steud. = *grossularioides* (L.) L'Hérit. 115
patulum Jacq. var. *patulum* 39, 46, 60, 62, **63–64**
patulum Jacq. var. *grandiflorum* N. van Wyk **64**
patulum Jacq. var. *tenuilobum* (Eckl. & Zeyh.) Harv. **64**
pedicellatum Sweet = *pulverulentum* Colv. ex Sweet 123
peltatum (L.) L'Hérit. 7, 26, 42, **49–50**, 52, 136, 149

pendulum (Sweet) G. Don = *longicaule* Jacq. var. *longicaule* 88
pillansii Salter = *pulverulentum* Colv. ex Sweet 123
pinnatum (L.) L'Hérit. *18*, 65, **68–69**, 139
plurisectum Salter 75, **82–83**, 126
polycephalum E. Mey. = a form of *carnosum* (L.) L'Hérit. 92
populifolium Eckl. & Zeyh. = *ribifolium* Eckl. & Zeyh. 109
praemorsum (Andr.) Dietr. subsp. *praemorsum* 28, 71, 72, **73**
praemorsum (Andr.) Dietr. subsp. *speciosum* Scheltema **73**
procumbens (Andr.) Pers. = *nanum* L'Hérit. 117
pseudofumarioides Knuth see under *minimum* (Cav.) Willd. **117**
pseudoglutinosum Knuth 40, **107–108**
pulchellum Sims 75, **83**, 85
pulverulentum Colv. ex Sweet 118, **123**, 124
punctatum (Andr.) Willd. **69**

quercetorum Agnew *23*, 71, 72, **73–74**, 133
quercifolium (L.f.) L'Hérit. *40*, 107, **108**
quercifolium (L.f.) L'Hérit. var. pinnatifidum L'Hérit. see under *quercifolium* 108
quinatum Sims = *praemorsum* (Andr.) Dietr. 73
quinquelobatum Hochst. 8, **50–51**

radens H.E. Moore 38, 95, 96, 100, 103, 104, **108–109**
radula (Cav.) L'Hérit. = *radens* H.E. Moore 109
radulifolium (Eckl. & Zeyh.) Steud. 122, **123–124**
ramosissimum auct. non Willd. see under *tragacanthoides* Burch. 86

ranunculophyllum (Eckl. & Zeyh.) Bak. see under *alchemilloides* (L.) L'Hérit. **44**
rapaceum (L.) L'Hérit. *19*, 66, **69–70**
rapaceum (L.) L'Hérit. var. *corydalifolium* Harv. 69
rapaceum (L.) L'Hérit. var. *luteum* Harv. 69
rapaceum (L.) L'Hérit. var. *selinum* Harv. **69**
relinquifolium N.E. Brown = *dichondrifolium* DC. 127
renifolium Swinbourne see under *inodorum* Willd. **116**
reniforme auct. non Curtis 53, 54, 59, **129–130**
rhodanthum Schltr. = *magenteum* J.J.A. van der Walt 59, 130
ribifolium Jacq. **109–110**
rodneyanum Mitchell ex Lindl. 26, **83–84**
rogersianum Knuth = *worcesterae* Knuth 86
roseum (Andr.) DC. = *incrassatum* (Andr.) Sims 67
roseum as cultivated by T. Moore see under *salmoneum* 51
'rotundipetalum' = a form of *carnosum* (L.) L'Hérit. 92

salmoneum R.A. Dyer **9**, 46, **51**, 52, 135
saniculaefolium Willd. see under *grandiflorum* (Andr.) Willd. 62
scabroide Knuth **110**
scabrum 'Apricot' 100
scabrum (Burm. f) L'Hérit. 98, **110**, 144
scabrum (Burm. f) L'Hérit. var. balsameum (Jacq.) Harv. = *scabrum* (Burm. f) L'Hérit. 110
scandens Ehrh. = form of *zonale* (L.) L'Hérit. 55
schizopetalum Sweet 49, **124**
scutatum Sweet = *P. peltatum* (L.) L'Hérit. 49
senecioides L'Hérit. **117–118**

sericifolium J.J.A. van der Walt 75, **84**
setiferum DC. = *elegans* (Andr.) Willd. 38
setulosum Turcz. 39, **40**, 75
sibthorpiifolium Harv. 18, **59–60**
sidaefolium (Thunb.) Knuth = *sidoides* DC. 130
sidoides DC. 54, **129**,130
sisoniifolium Baker = *carnosum* (L.) L'Hérit. 92
spinosum Willd. 27, 75, 82, **84**
spondylifolium Salis. = *lobatum* (Burm. f.) L'Hérit. 121
stenopetalum Ehrh. 9, **52**
stipulaceum (L.f.) Willd. 28, 75, **84–85**
sublignosum Knuth 41, **110–111**
subherbaceum Knuth = *multicaule* Jacq. subsp. *subherbaceum* (Knuth) J.J.A. van der Walt 89
suburbanum Clifford ex Boucher subsp. *bipinnatifidum* (Harv.) Boucher **90**
suburbanum Clifford ex Boucher subsp. *suburbanum* 31, **90**
synnoti (Sweet) G.Don = *myrrhifolium* (L.) Hérit. var. *myrrhifolium* 89

tabulare (L.) L'Hérit. = *elongatum* (Cav.) Salisb. 45, 64
tabulare (Burm. f.) auct. non L'Hérit. 16
tenuicaule Knuth 75, 85
tenuifolium L'Hérit. = *hirtum* (Burm. f.) Jacq. 80
terebinthinaceum (Cav.) Desf. = 'Graveolens' 103
ternatum (L.f.) Jacq. 61, 62, **64**
tetragonum (L.f.) L'Hérit. 22, 71, **74**, 87
tomentosum Jacq. 41, **111**, 136
tongaense Vorster **52–53**
tragacanthoides Burch. 75, 80, **85–86**
transvaalense Knuth 53
trichostemon Jacq. = *ovale* (Burm. f.) L'Hérit. **40**
tricolor Curt. 2, 3, 36, 38, 39, 40, **41**

tricolor Curt. var concolor Harv. = *capillare* (Cav.) Willd. 37
tricolor var. arborea as found in gardens = 'Splendide' 41
trifidum Jacq. 29, 75, 79, **86**
tripartitum Willd. = *trifidum* Jacq. 86
trifoliolatum (Eckl. & Zeyh.) E.M. Marais see under *pinnatum* (L.) L'Hérit. **69**
trilobatum sensu Eckl. & Zeyh. = *scabrum* (Burm. f.) L'Hérit. 110
triste (L.) L'Hérit. 15, 25, 66, 118, 119, 122, 123, 124–125

urbanum (Eckl. & Zeyh.) Steud. = *suburbanum* Clifford ex Boucher subsp. *suburbanum* 90
urbanum (Eckl. & Zeyh). Steud. var bipinnatifidum Harv. = *suburbanum* Clifford ex Boucher subsp. *bipinnatifidum* (Harv.) Boucher 90
usambarense Engl. see under *multibracteatum* Hochst. **48**

variegatum (L.f.) Willd. see under *grandiflorum* (Andr.) Willd. 62
veronicaefolium (Eckl. & Zeyh.) Steud. = *ovale* subsp. *veronicifolium* (Eckl. & Zeyh.) Hugo 40
viciifolium DC. see under *pinnatum* (L.) L'Hérit. **69**
violaceum of gardens = *tricolor* Curt. 41
violareum Jacq. = *tricolor* Curt. 41
vitifolium (L.) L'Hérit. 95, 97, 98, 107, **111–112**

whytei Baker see under *multicaule* Jacq. **89**
woodii N.E. Brown = *schizopetalum* Sweet 124
worcesterae Knuth 75, 81, **86**
xerophyton Schltr. ex Knuth *12*, 56, 58, **60**

zeyheri Harv. = *luridum* (Andr.) Sweet 122
zonale (L.) L'Hérit. **10**, *10*, 17, 26, 42, 46, 51, **53–55**

× *domesticum* Bailey 99
× *hortorum* Bailey 46
× *kewense* R.A. Dyer **55**
× *tricuspidatum* L'Hérit. **110**

'Ardens' 79, **122**
'Asperum' **102**
'Atomic Snowflake' **98**
'Attar of Roses' **98**
'Clarissa' 51
'Fernaefolium' = 'Filicifolium' 101
'Filicifolium' **101**
'Formanii' 62
'Graveolens' **11**, *38*, 81, 136
'Islington Peppermint' **41**
'Karooense' 81
'Lady Plymouth' **103**
'Lateripes' **50**
'Lavender Lad' **128**
'Lavender Lass' **128**
'Lawrenceanum' **122**
'Lilac Lady' **128**
'Little Gem' **103**
'Mabel Grey' **98**
'Miss Stapleton' 56, **58**, 59
'Monsieur Nonin' 110
'Mrs Kingsbury' 69
'Nobilis' 62
'Otto of Roses' 98
'Paton's Unique' 110
'Pheasant's Foot' **102**
'Pink Quink' **51**
'Purple Unique' 100
'Radula' 103, **109**, 136
'Rosé' 103, **104**, 109
'Royal Oak' *40*, **108**
'Salmonia' = 'The Boar' **46**, 51
'Saxifragoides' **51–52**

'Scarlet Unique' **79**
'Scandens' **55**
'Schottii' **122**

'Splendide' 3, 36, 40, **41**
'The Boar' **46**, 51
'Viscossisimum' **102**
'White Boar' **46**

GENERAL INDEX

Andrews, H. 18, 19, **22–23**

Banks, Sir J. 17, 18
Bentinck, W. 16
Boos, F. 17
Breyne, J. 16, 90
British East India Company 15
Burmann, J. 17
Burmann, N 17

Candolle, A.P. de 27, 29
Cavanilles, A. 17, 26
Commelin, C. 16, 17
Commelin, J. 16
Compton, H. 16
Cook, Captain J. 18
Cultivation **133–136**
  containers 134
  growing media 134
  humidity 133
  light 132
  pruning 135
  temperature **133–134**
  uses in the garden **135–136**
  water **132–133**

Dillenius, J. 16, 26, 51
Diseases **147–148**
  bacterial blight 147
  black leg 147
  galls 147
  grey mould 148
  rust 148
  sooty mould 148
  virus 27

Don, G. 27
Dutch East India Company 15, 16, 18

Gardens
  Amsterdam 16, 17
  Chelsea Physic 16
  Chiswick 27, 52
  Kew 17, 18, 41, 52
  Leiden 15, 16, 17
  Oxford 16
  Schönbrunn 17, 36
Gordon, Captain R. 18

Harvey, W. 27, 29
Hermann, P. **15–16**
Hoare, Sir R. Colt 19, 64
Hove, A.P. 18

Jacquin, N. 17, 36
Jenkinson, R. 19, 70

Knuth, R. 27

L.'Héritier 17, **22–23**
Linnaeus, C. 17, 23, 26, 48
Lindley, Dr. J. 27

Masson, F. 17
Morin, R. 15

Nurseries
  James Colvill 18
  Andrew Henderson (Pine Apple Place) 18
  Lee and Kennedy (Vineyard Nursery) 18

Oldenland, H. 16
Parkinson, J. 25
Paterson, W. 18

Pelargoniums
  adaptation to climate **11–12**, 30, 36, 56, 60, 65, 71, 75, 87, 91, 112, 118, 125
  characteristics **23–25**
  classification **11–12**, 21–28
  distribution 11
  diversity of habit 10
  hybridization 74, 95, **141–142**
  hybrids **12–13**, 19, 36, 40, 42, 51, 59, 79, 97, 100, 107, 110, 121, 122, 130
  ivy-leaved 10, 19, 42, 50
  medicinal uses 43, 53, 65, 72, 97, 117, 120, 123, 125, 129, 130
  modern taxonomic methods 27
  oils 10, 96, 104, 109, 111
  regal 10, 19, 95, 97, 100
  scented 10, 19, 95
  unique 10, 19, 95
  zonal 10, 19, 42, 46, 55
Pests **143–147**
  aphids 145
  biological control 144, 146, 147
  caterpillars 145
  mealy bugs **145–146**
  red spider mites 146
  slugs and snails 146
  vine weevils 146
  whitefly **146–147**
Physiological Disorders **148–149**
  fasciation **148–149**
  oedema 149
  wilting 149
  yellowing leaves 149
Propagation **137–142**
  division **139–140**
  root cuttings **139–140**
  seed **138–139**, **140–141**
  stem cuttings **138–139**

Royal Horticultural society 19, 27, 41, 46, 51, 52, 73

Scholl, G. 17
Sherard, J. 16
Stel, S. van der 16
Sweet, R. 18, 19, 27, 29, 52, 64, 70

Thunberg, C. 17
Tradescant, J. 15

Vorster, P.J. 28

Walt, J.J.A. van der 28

■ SOURCES OF INFORMATION

Collections of *pelargonium* species in cultivation

Chelsea Physic Garden
66 Royal Hospital Road
London
UK

Conservatoire National du Pelargonium
Bourges
France

Geelong Botanic Gardens
Geelong
Victoria
Australia

Institut für Botanik
Schloss Garten
Munster
Germany

National Botanical Gardens of South Africa
Kirstenbosch
Claremont
Republic of South Africa

National Botanical Gardens of South Africa
Worcester
Republic of South Africa

National Council for the Conservation of Plants and Gardens (National Collection of Pelargoniums at Fibrex Nurseries)
Pebworth
Warwickshire
UK

New York Botanical Gardens
Bronx
New York
USA

Royal Botanic Gardens
Kew
near Richmond
Surrey
UK

Royal Botanic Gardens
Sydney
New South Wales
Australia

Specialist libraries

The following libraries are not all freely accessible to the public and it may be necessary to make an appointment before visiting.

Hunt Institute of Botanical Documentation
Pittsburg
Pennsylvania
USA

The Lindley Library
The Royal Horticultural Society
80 Vincent Square
London
UK

National Botanical Gardens of South Africa
Kirstenbosch
Claremont
Republic of South Africa

National Botanical Gardens of South Africa
Kirstenbosch
Claremont
Republic of South Africa

New York Botanical Gardens
Bronx
New York
USA

Specialist societies

Australian Geranium Society
British and European Geranium Society
British Pelargonium and Geranium Society
Geraniaceae Group (UK)
International Geranium Society
Italian Pelargonium Society

Sources for *Pelargonium* species

The RHS Plant Finder, published annually, and widely available in book shops and garden centres, lists nurseries in the UK and the Republic of Ireland, and the species and cultivars they supply. The European Plant Finder includes nurseries from nine European countries including France, Germany and the Netherlands. It is also available as a CD Rom. There are also individual plant finders in several other countries which may be helpful.